SAVING
AMERICA'S BEACHES

ADVANCED SERIES ON OCEAN ENGINEERING

Series Editor-in-Chief
Philip L- F Liu (*Cornell University*)

Advanced Series on Ocean Engineering — Volume 19

SAVING AMERICA'S BEACHES

The Causes of and Solutions to Beach Erosion

Scott L. Douglass

University of South Alabama
USA

World Scientific
New Jersey • Singapore • London • Hong Kong

Published by

World Scientific Publishing Co. Pte. Ltd.

5 Toh Tuck Link, Singapore 596224

USA office: Suite 202, 1060 Main Street, River Edge, NJ 07661

UK office: 57 Shelton Street, Covent Garden, London WC2H 9HE

British Library Cataloguing-in-Publication Data

A catalogue record for this book is available from the British Library.

First published 2002
Reprinted 2003

ISBN   981-238-097-3 (pbk)

Printed in Singapore by Mainland Press

Dedicated to

my wife and children,

and to all other beach lovers.

Preface

Saving America's Beaches

Americans love beaches. From the Jersey shore to Santa Monica Bay, sandy beaches are part of the fabric of society. Millions want to vacation at the beach and millions want to live near the beach. The beach is America's longest playground! Yet, it appears to be in worse shape than ever before. More beaches are losing sand than are gaining sand and they are going to be even more threatened by tremendous natural and human forces in the coming years.

Many different solutions have been proposed. One of the best ways to avoid a problem with erosion is by "backing off" the beach. Building back away from the water behind a "setback" allows the beach to come and go without erosion becoming a problem. However, the temptations of building close to the sea have been too great for many of America's most popular beach resorts. We've already decided to live and to play on barrier islands and coastal bluffs. And we've already built houses and condominiums there. It is now only a matter of how we are going to live there. Other solutions are needed for the survival of healthy beaches. This book sheds light on two critically important issues related to how we live there: one, the cause of much beach erosion and, two, the debate about beach nourishment as a solution.

One of the little known truths about beach erosion in this country is that much of it is man-made! We've destroyed our own beaches. While there are natural reasons for beaches to erode, most of the worst problems are not natural. Scientists have found that an awful lot of beach erosion is caused by "works of man" such as ship channels, coastal structures including jetties and groins, dams and flood control works on the west coast, and mining of beach sand. For example, millions of cubic yards of beach sands have been dredged from boat and ship channels but not placed back on the beaches from which the sands came. Sometimes, the beach sands are dumped in deepwater offshore - lost to the beaches forever. This starves the downdrift beaches of the sand that is their very lifeblood. Blocking and intercepting sand moving along the coast causes terrible beach erosion in this country.

We have solutions to beach erosion that work. One solution that works is restoring the natural movement of sand along and to our beaches. Dredged sand can be artificially "bypassed" around the ship channels to the downdrift beaches. However, we have done a lousy job of bypassing sand in America and our beaches have suffered the consequences - erosion.

A second little-known truth concerns beach nourishment, the direct placement of large amounts of good sand to widen the beach. Beach nourishment is widely criticized as unnatural, expensive, and ultimately futile because it washes away. It has even been called "the fleecing of America." But surprisingly, beach nourishment has worked well! Hundreds of miles of America's beaches, including many of our most popular beaches, have been saved by beach nourishment. This book examines the criticisms and concludes that beach nourishment is more natural than the most common realistic alternative of a seawall, is not expensive compared with the value of the beach, and really doesn't wash away when done correctly except to wash down the coast to feed other beaches! On many beaches, nourishment is replacing the sand removed by other works of man. It is often the best of the available approaches to saving beaches.

We have plenty of experience with other approaches to beach erosion that do not save sandy beaches. We have centuries of experience with seawalls of many types that protect upland property but do not save the sandy beach. "Armoring" an eroding beach with a seawall may save a building but destroy the sandy beach we love. We also have plenty of experience with different forms of so-called "experimental" solutions. Many of these might be called "snake-oil" since they really aren't based on sound science and a proven track record but rather on slick sales pitches to desperate consumers. At best, these experiments are a distraction for those looking for good solutions. Some proposed public policy solutions are aimed at discouraging coastal development by raising flood insurance premiums, stopping public funding for roads and other infrastructure, and stopping mortgage lending near the beach. While such policies may be appropriate, they seem to ignore several facts: one, many of our beach erosion problems are caused by poor sand management; two, alternative engineering solutions such as sand bypassing and beach nourishment work and can accomplish many of the same goals cheaper; and three, the sandy beach has tremendous value to our society. Such policies are not going to help the sand-starved beaches and they may needlessly make it harder for millions of Americans to enjoy sandy beaches.

Good policy is based on the will of an informed public. America should make future beach policy choices based on a clear understanding of the causes of and solutions to beach erosion.   This book describes how beaches work; where the sand comes from, how waves are formed and how they break, how this moves sand down the coast, how man has interrupted this movement; and what needs to be done to save our beaches for future generations of Americans. So if you love waves and beaches, and you care about the future of your favorite beach spot, then read this book while you enjoy the beach!

Contents

Chapter 1

Beaches — America's Longest Playgrounds

Sometime during the first week in July, on a steaming hot beach in America, 'Little Ray' Hogan is going to take a cup of ice water and creep up on his big sister Penny, while she sleeps in a beach chair, with about a dozen smirking relatives watching, and – you guessed it – dribble some of the ice-water on his big sister. And when Penny jumps up screaming "Raymond!" and he scampers away safely, the on-looking relatives will all laugh and hoot and holler and have a good time. A few of them have watched 'Little Ray' do this to his big sis for over 50 years! And when he does it now, the children, and grandchildren and nephews and nieces are part of the weeklong vacation. They come from different cities and states to gather together at the beach to talk, and eat, and fish, and golf, and surf, and visit the arcades, and do puzzles, and swim, and eat some more. But mainly they talk, about teenagers, and health, and old friends and loved ones. That is America. And it is part of the quality of life in America.

Access to healthy beaches improves our quality of life. Millions of Americans choose to spend some of their limited vacation time at the beach.

Beaches are some of America's greatest economic and environmental assets. Beaches are America's longest playgrounds! Beaches provide places to play and relax that are unmatched. Every coastal state has great beaches for its citizens and visitors.

Many beaches are easy to get to and to get on. Over 500 million visits are made to beaches in California every year. And there are a myriad of things to do at a beach.

Walking the shoreline is a favorite pastime at almost every beach in America. Whether it's a footpath in a grassy pasture above a rocky shoreline or a sandy, low tide beach on a barrier island, there is something soothing to the soul about walking along the edge of the land with an endless sea of waves breaking beside you. "Beachcombing" or "sea shelling" is a beach walk to observe shells, driftwood and flotsam and jetsam. Dozens of great books have been written on seashells. Each portion of the American coast has different, fascinating seashells that are the remains of the local, natural beach life that develop hard shells.

Walking along a sandy beach is soothing to the soul.

Probably the all-time favorite thing to do at the beach is absolutely nothing! There are many different ways to do nothing at a beach. "Laying-out" or "working on your tan" are favorites of the younger crowd in spite of the dangers to the skin. Reading books and newspapers is a favorite of many beachgoers. Publishers release and market special types of "beach books" for those of us that prefer to relax our minds while at the beach.

The list of ways to play at the beach includes flying kites, playing ball games of all sorts including the unique sport of beach volleyball, building sand-castles, gazing at stars at night, whale-watching, people-watching, fishing, pumping iron at "muscle beach" in Venice Beach California, running, biking, playing cards, sleeping, and on and on. And all of these activities are on the sandy beach. A list that is just as long is needed to list the way people play in the water at the beach! This includes swimming, surfing, "fanny-dunking," raft-riding, boogie boarding, body-surfing, sailing, skim-boarding, wind-surfing, kayaking, rowing, jet-skiing, water skiing, and kite-surfing.

Beaches can be thought of as America's longest playgrounds! We play on them in many ways.

Some beaches have rather unique ways to play. Driving cars on the beach is a form of recreation at several American beaches. The original Daytona 500 car races were run on the beach. Daytona; Galveston, Texas; Pismo Beach, California; and Long Beach, Washington are just of few of the coasts with a tradition of beach driving for recreation that continues. Many other beach areas allow limited beach driving for local surf fishermen. Clamming is a tremendously popular activity along some American beaches. Over 90,000 people have been counted along one stretch of beach in Washington during the "clamming season."

Beaches are economic engines

The overall social and emotional value of beaches may be tough to measure but the economic facts are clear. Beaches are tremendous economic engines. Millions of Americans want to vacation at the beach and millions of Americans want to live near the beach. Jim Houston, Director of the US Army Engineer Research and Development Center, has gathered many of the numbers about the economic impact of beaches. His arguments are briefly outlined here.

According to the World Travel and Tourism Council, travel and tourism is the world's largest industry. It is also one of America's largest industries. It is estimated that 17 million American jobs, almost one out of every eight, is in the travel and tourism portion of the service sector of the economy. Travel and tourism provides $1.2 trillion, or 12%, of America's Gross Domestic Product.

Clearly, travel and tourism is a large portion of the American economy. And it is growing. The service sector has been growing and is projected to continue to provide most of the job growth in America. Some economists have found that travel and tourism jobs pay higher average wages than the U.S. per capita wages. Thus the perception that most of these jobs are "low-paying" may not be correct.

Healthy beaches provide hundreds of thousands of American jobs! Beaches are the linchpins of the American tourism industry.

Travel and tourism is one American industry with a very large, consistent foreign trade surplus. In 1999, that surplus was almost $14 billion according to the International Trade Administration and Bureau of Economic Analysis. In other words, foreigners spend more money traveling in America than Americans spend traveling overseas. Thus, healthy beaches are good for the national economy as well as local and regional economies in the coastal areas.

Beaches are the linchpins of the American tourism industry. Many preference polls have found that beaches are the number one destination for tourists. And the numbers support those preferences. There are more visitors to one beach, Miami Beach, than the 3 largest national parks combined (Yellowstone, Grand Canyon and Yosemite).

Healthy beaches are directly responsible for hundreds of thousands of American jobs. Bill Stronge, economics professor at Florida Atlantic University, estimates that almost 400,000 jobs and over $8 billion in payroll result from additional spending in Florida due to the state's beautiful beaches. About 25%, or over $25 billion, of the coastal real estate value in Florida can be attributed to the state's beautiful beaches. Beaches are valuable to all states. Other studies show that California beaches provide over 800,000 jobs with $14 billion in direct revenue and New Jersey's beaches provide 270,000 jobs with $6 billion in wages.

Billions of dollars in tax revenues come from beach tourism to federal, state, and local governments. One former Florida politician used to say that every dollar that the State of Florida spent on its beaches came back 12 times in increased private business revenues and thus increased tax revenues.

The federal government appears to gain the most from beach tourism. Economic studies consistently show that the federal government receives most of the tax benefits from beach tourism. Income taxes come from the millions of Americans employed by beach tourism. Local and state governments also benefit from increased sales taxes and property taxes.

Every coastal state has beaches for the public to visit. There are over 500 million "beach visits" to the California coast each year. Beaches are part of the quality of life as well as a big part of the economy.

The point is - anything that hurts the beach also hurts the economy. Disposing of sand dredged from ship channels anywhere else but on the beach hurts the economy. Trapping sand that was on its way along the coast with coastal structures hurts the economy when erosion is caused elsewhere. And mining sand from the beaches, inlets, and river mouths removes sand from the beaches and hurts the economy. Each of these activities contributes to beach erosion because the health of the sandy beach is reduced.

Beach nourishment, on the other hand, adds sand to the beaches. The addition of sand can save the recreational and aesthetic aspects of the beach that control its economic value as well as provide storm protection. Jim Houston points out that our foreign trade surplus in tourism has been shrinking in the last decade and concludes that spending money to restore and maintain beaches is needed as a national infrastructure investment. Without spending some money, "the U.S. will relinquish a dominant worldwide lead in its most important activity."

Beaches are environmental treasures

Healthy beaches are also unique environmental treasures. Many plants and animals use the beach and some; including endangered turtles, beach mice and plovers; require it for survival. For them, the saying "life is a beach" is absolutely true! Sea turtles are hatched in nests buried in beach sands along the southeastern coast of the United States. The nests are buried above the high tide line at the base of sand dunes.

Many plants and animals require a sandy beach for survival. These shorebirds feed in the swash zone, the part of the beach where the waves swash back and forth across the sand.

The mothers crawl out of the sea at night, dig into the sand, lay their eggs, and then crawl back to the sea. Without a sandy beach, there is no place for these nests.

There are several unique habitats on beaches. Shorebirds feed on the tiny organisms in the swash zone as each wave stirs them up. Different types of crabs live in the dry and wet parts of a beach. Beach mice inhabit many American sand dunes. Many plants, including beach and dune grasses, thrive in this dynamic environment. Storms, which occasionally bring high water and waves to the dunes, control the physical aspects of the dunes and upper beach. The physical aspects, in turn, control the biological aspects of the beach and dune ecosystem.

Beaches are important to America is so many ways: emotionally, economically, and environmentally. We need to make wise decisions about the future of America's beaches.

Sand dunes, like these on South Padre Island Texas, provide unique habitat for plants and animals as well as protection from storms.

Sea lions resting on a California pocket beach.

Chapter 2

Our Jeweled Necklace of Sand — The Geology of Beaches

> The surf was booming loudly as Bob and JoeAnn walked along the beach. They walked or ran on this beach everyday when the weather was pleasant. The sand, the surf, and the smell of the salt in the wind somehow always seemed fresh and different. But many of the same forces were working today as they have worked for years. The sandy beach is the product of thousands of years of geology influenced by decades of human impacts.

America is defined in many ways - one way is geography. The nation stretches from the Atlantic coast to the Pacific coast and from the shores of the Great Lakes to the Gulf of Mexico beaches. Some of these shorelines are rock cliffs that extend straight down into the sea and are pounded by relentless ocean waves. Some of these shorelines are muddy with almost no waves such as the mangrove coast of the Everglades in southwest Florida. But, most of America's shorelines are sandy beaches.

All beaches are not alike! Beaches, the accumulations of loose sediments along the shoreline, are very different throughout America. Some beaches are long and straight: some beaches are short and curved "pocket" beaches between rocky headlands. Some beaches have sand: some beaches have rocks. Some beaches are backed by high cliffs: some beaches have nothing behind them but sand dunes on a barrier island just a few feet above sea level. Some beaches have bright white sand: some have black sand. And most beaches have sand that is somewhere between light tan and brown with many different hues and shades.

The "look" of every beach today is partly the product of thousands of years of geology at work and partly the product of decades of human impacts. Many of the variations in beaches are due to different geologic histories. Understanding this history, the so-called "geologic framework," that makes each beach the way it is today is important to understanding how to save our beaches. This chapter explains the ancient geology of beaches while the next two chapters explain the impacts of waves and humans on our beaches today.

Mountains are often close to the sea on the Pacific coast. In the Big Sur region they come right down to the sea.

Beach sands

The sand on most beaches is whatever hard, loose sediments are available, based on the local geology, at that location. Sand at the base of an eroding bluff usually matches the sand in the bluff. But many of America's beaches are sand that was moved there by waves. The size of individual grains of beach sand varies from about 0.1 mm to 1 mm. Waves typically remove the smaller grains, silt and clay particles, from the beach. And because of the better mobility of sands, gravels are much less common on beaches than sands. The exceptions are the many "cobble" beaches of New England, the Pacific and the Great Lakes. Layers of rocks on a beach are called "cobbles" if the rocks are round like a ball and called "shingles" if they are flattened such that they are good for skipping across the water.

Some American beaches have a layer of "cobbles," round rocks, on top of beach sand like this San Onofre, California beach.

Some of the whitest beaches in America are found on the Gulf of Mexico coast of Florida. These beach sands are quartz minerals that washed out of the ancient southern Appalachian Mountains hundreds of thousands of years ago. Because of the sorting that has occurred since, they are now almost 100% quartz sand grains of almost all the same size.

Some beach sands are mostly broken-up seashells or corals. They were originally formed from carbonate minerals by animals for protection. When the animal dies, the shell becomes part of the beach system. The natural rates of carbonate "shell hash" production are so great on some beaches it is the dominant loose sediment that makes up the beach sand. South Florida has some beaches that are almost entirely carbonate sands with little or no quartz sands. While there are some volcanic-sand beaches in Hawaii, most of those beautiful beaches are made of carbonate sands. Shell-hash on beaches is found throughout America.

The size of the sand grains influences the way a beach behaves naturally and how Americans use it. The larger the grains of sand, the steeper the beach slope: the finer the

American beach sands come in a wide variety of natural colors. Some samples are shown here. These are from: top row; Presque Isle, PA; Long Beach, NY; and Myrtle Beach, SC; second row; Daytona Beach, FL; Ocean City, MD; San Clemente, CA; Long Branch, NJ; and Manasota Key, FL; third row; Long Beach, WA; Headland Beach State Park, OH; Olympic National Park, WA; Flagler Beach, FL; and Oregon Dunes, OR; fourth row; Big Sur, CA; Pacifica, CA; Walton County, FL; Ocean City, NJ; and Siesta Key, FL.

grains of sand, the milder the beach slope. Median grain size is often used to characterize the overall beach sand. On any beach, there is always some variation in grain size about this median. The larger grains are found where the waves break and the smaller grains are found offshore and in the sand dunes. Median grain sizes on America's beaches vary from about 0.15 mm to 0.8 mm. This doesn't seem like much but the difference can be very noticeable. Smaller-grained sands seem to blow around on the beach more readily and get in one's hair and clothing more easily! Coarser grained sands seem to be harder to walk across because your feet sink in farther.

The ease of both sandcastle building and driving on the beach are related to beach sand grain size. Surface tension forces between water and sand grains are stronger with finer sand grains that pack together better. Thus, sandcastles made of finer sands can often hold more ornate designs. Also, because the finer sands pack down tighter, driving cars on the beach is easier. The places with a long tradition of extensive beach driving, such as Texas and Washington, are those with finer beach sand grains.

Some beach activities are related to special, almost unique, local aspects of the beach geology. Searching for ancient shark's teeth on the beach is a popular pastime along a few American beaches. Some of the interpretive staff at the Hunting Island State Park in South Carolina enjoy startling people by walking along the beach until they stop, mention that if one looks very carefully they might find some shark's teeth, and then casually stooping over to pick one up! Although it might appear to be sleight of hand, it really is just experience in knowing where, when and how to look. The fossilized teeth are from sharks that became extinct millions of years ago. Storm waves wash them to the beach regularly. The beaches of Manasota Key, on Florida's Gulf coast, produce so many ancient shark's teeth, that the local beachcombers have developed a device they call a "Florida snowshovel." It is a wire basket on a stick used to sift the sand along the water's edge for the shark's teeth.

This retiree traded in his snow shovel for this "Florida snow shovel," a wire basket on a stick used for finding ancient shark's teeth in the beach sands on the Gulf of Mexico beaches of Manasota Key.

Most sandy beaches are just a thin veneer of sand overlying some other geologic formation. The underlying rock formation is exposed on the eroding Florida beach shown above. These beach rocks provide a tremendous "reef" habitat where they stick out of the sand underwater just offshore. The erosion of the thin veneer of sand on the sandy barrier island beach shown below has exposed some underlying peat from an old marsh and some tree trunks from an old forest. Peat outcroppings like this are not unusual on eroding Atlantic and Gulf of Mexico beaches.

This cobble beach on the Olympic Peninsula in Washington "gurgles" like a waterfall with each receding wave as water seeps into and tumbles down through the rocks.

Grain size is also important when placing sand on a beach for sand bypassing or beach nourishment. Nourishment works better, physically and environmentally, when the sands match the native beach sands as closely as possible.

One of the most surprising, and important, aspects of beaches is how thin they are. Beach sands are a very thin veneer overlaying the ancient, local geology at most places. For example; on many Pacific, Great Lakes, and New England beaches; the beach sands lay on a shelf of bedrock. Even on the sandy barrier islands of the Atlantic and Gulf coasts, the beach sands usually lay on top of layers of rocks or mud. Underwater outcroppings of rocks or mud are common offshore of almost every beach in America. Rock outcroppings are sometimes called reefs because of their shallowness and the tremendous diversity of marine life they attract. They are commonly seen off the California coast because they affect the waves so much. They also are also common in warm water. In Florida, for example, there are reef rock outcroppings in the nearshore and even occasionally on the beach.

The thinness of the beach sand deposits is one reason to be sure that man-made sand removals are minimized. For example, rock out-croppings on the beach and in the surf are more common downdrift of inlets on the Atlantic coast of Florida. Erosion caused by man's activities at the inlet has removed the thin veneer of sand and exposed the rocks. Often the sand has been trapped by the inlet

Barrier islands are sandy islands separated from the mainland. The Outer Banks of North Carolina are an extreme example of this. Many of America's most beautiful and productive estuaries are in the lee of barrier islands.

structures or dumped offshore by dredging.

Mike Chrzastowski of the Illinois State Geological Survey says that if you were to drain the Great Lakes, the sandy beach would be a thin ring around the tub. The same is true of the Atlantic, Pacific and Gulf. There just is not an infinite amount of sand on most American beaches. Coastal managers and coastal geologists throughout the country say that they have a "sand-starved" system. And they are right whether they are talking about the pocket beaches of southern California, the eroding bluffs of Lake Erie or even the barrier islands of South Carolina.

The origin of sand is the weathering of mountains. But there is an important difference between the sandy barrier islands of the Atlantic or Gulf and the pocket beaches of the Pacific or New England. Sand from weathering and erosion in the canyons of the coastal mountains feeds many of the beaches of the Pacific today. But this probably has not been true for the beaches of the Atlantic for well over ten thousand years. The sands that make up many of the Pacific beaches coast washed out of the

Barrier islands can be extremely dynamic. This curved portion of Dauphin Island, Alabama on the Gulf of Mexico had a perfectly straight beach when surveyed in 1852 by Benjamin Franklin's grandson, A.D. Bache. The island became curved when the middle portion was overwashed by storms frequently in the early 1900's and it migrated over 1000 feet to the north (to the left in picture).

mountains within the last several thousand years. Some of it washed out onto the beach just a few years ago in the El Niño storms. By contrast, sands that make up most of the barrier island beaches of the Atlantic and Gulf of Mexico came out of the mountains

This barrier island, Hunting Island South Carolina, has eroded so far that the surf has moved into and is killing a maritime forest that used to be hundreds of yards inland.

hundreds of thousands of years ago and have been moved around by waves as sea level has fluctuated up and down since then. There is no more additional sand being added to the beaches of the Atlantic barrier island systems from the rivers today. The Atlantic coast of the United States has a very broad coastal plain between the mountains and the beaches and a wide continental slope offshore. Thus, the "gradient" or slope of the earth from the mountains to the sea is very mild. The small amount of sand and gravel eroding from the Appalachian Mountains today doesn't reach the sea because the river slopes are too gentle and the flow too slow.

Many barrier islands have provided recreational opportunities for millions of Americans for decades. This barrier island, the Wildwoods of New Jersey in 1985, has been a playground for generations of Philadelphians. It has one of America's great boardwalks along the sea.

Beach geology is like a giant four-dimensional jigsaw puzzle. The fourth dimension is time. The coasts are all changing with time. But beware! Geologists treat time differently than most Americans. For most of us, a human lifetime is a long time. But for geologists, ten thousand years is quicker than the blink-of-an-eye.

When dealing with the time scales that geologists find in the rock record around the world, millions and billions of years are needed. The earth was formed about 4 to 5 billion years ago and all of the geologic time periods and eras that we memorized in high school began about 500 million years ago with the Cambrian Period. To appreciate the length of these time scales, consider the yardstick analogy. If the entire history of the earth were laid out on a wooden yardstick from left to right, one swipe along the right edge with a nail file would remove all traces of humanity!

Sea level changes

Sea level changes are an important part of the beach geology puzzle. It appears that sea level has fluctuated tremendously throughout the past 2 million years and that we are now in a period of rising sea levels that began only about 15,000 to 20,000 years ago. The evidence for the geologists' estimates of ancient sea levels comes from a variety of sources including sediment and glacier drilling records and the geologic maps of the world. The ages of different layers in the drilling records are estimated through several sophisticated chemical analysis techniques. Ancient fluctuations in sea level have been related to global climate changes and the amount of water stored in the polar ice caps. During warm periods, the ice caps melted and sea level rose. During the ice ages, the ice caps and glaciers grew and sea level fell. Of particular importance is the past 20,000 years.

The worldwide-average sea level was probably about 300 to 400 feet lower 20,000 years ago than it is today. It then rose very rapidly until about 5,000 years ago when it almost stopped. While there have been some fluctuations in the past 5,000 years, the evidence indicates that the rapid rate of rise slowed dramatically. The important part is that sea level has been more or less constant, relative to other geologic times, for the past 4,000 years. This is the time period of recorded human history. The beaches that we

Low tide exposes extensive sand shoals offshore of some inlets like this one in South Carolina. Sand often moves from these shoals to the beaches and back.

Curved pocket beaches are held by resistant headlands. These two pocket beaches are at the base of the eroding bluff in Palos Verdes, California.

Leadbetter Beach in Santa Barbara, California is a pocket beach between a natural headland and an artificial headland. The artificial headland is the breakwater at the port of Santa Barbara in the background and this photo was taken from the natural headland bluff.

Inter-tidal rocks and pools where the sand veneer is swept away, like this area in Laguna Beach, California, provide exceptional habitat for marine life.

have today along our coasts have been influenced by this pattern of sea level changes. When sea level is more or less constant, waves have the time to move tremendous amounts of sand along the beach and even perhaps across the continental shelf. The barrier islands on the Atlantic coast have probably formed during the last 5,000 years. The sand that washed out of the Appalachian Mountains and deposited on the coastal plain and continental shelf many hundreds of thousands of years ago just got reworked during the past 20,000 years as sea level rose over the ancient deposits. During the past 5,000 years of relatively similar sea levels, the sand has moved along the coast (and maybe even to the beaches across the continental shelf) driven by waves to form the barrier islands.

Coastlines were farther offshore than they are today. During that last "low-stand" in sea level, roughly 20,000 years ago, the Atlantic coast was up to a hundred miles offshore of today's beaches. The distance was not as great on the Pacific coast because the continental shelf is much narrower and deeper.

For the past century, tide gauges have been able to measure relative sea level rise around the country. This is the net change between the land elevation and the water elevation. Most land around the country is either rising or falling slowly because of some local geology. For example, some beaches in Texas, Louisiana, and California are subsiding because of compaction and extraction of hydrocarbons by oil wells. The land around much of the Great Lakes is still rising, or rebounding, from the last glacial retreat because the weight of the ice was removed. Along the Pacific coast, which is tectonically much more active, some parts of the land mass are rising faster than sea level is rising. There is a relative sea level fall at these places!

"Stacks," such as these above along the Oregon coast, are fairly common along the Pacific coast of America. They are harder, more resistant rocks that used to be part of the coastal bluff until the bluff eroded past them. "Arches" or "bridges" such as this one in Santa Cruz, California below eventually will collapse leaving two stacks.

The beach shown above, at Point Reyes National Seashore in California, is backed by a high, eroding bluff that provides the sand for the beach. The brightly colored clay layers in the bluff shown below, at Gay Head on Martha's Vineyard, Massachusetts, are kept bright by erosion by rain and by waves at the base of the bluff.

Whether the relative sea level is rising or falling depends on where you are. For most of the Atlantic and Gulf coasts, there is a relative rise in sea level of 6 to 10 inches in the last century. The rate is much greater than this for Louisiana's coast. On the Pacific coast, there is no overall pattern. For example, at San Francisco there has been about a 6-inch rise. However, several hundred miles north, near the California/Oregon border, there has been a 6-inch drop.

The worldwide, average rate of sea level rise, with land changes subtracted out, is probably about 1 to 2 mm/year for the past one hundred years. Scientists have looked for, but not found (at least not yet), a global increase in sea level rise rates in the past few decades. Future sea levels are going to be critical. Many atmospheric scientists have concluded that the earth is warming now and that sea level rise rates will accelerate in response. Of course there is, right now, significant uncertainty about future sea level rise rates. Will the rate increase a little or a lot? And how will the beaches, which have been able to respond to the current rate of sea level rise, be able to respond to accelerated rates of rise?

If there are increases in the rate of sea level rise in the next century, we can be sure that our coastlines will feel it. Keeping sand on the beaches may be even more important for the future of America's sandy beaches. Eliminating the avoidable sand losses caused by man should be high on our agenda right now.

Chapter 3

Surf's Up! — Waves and Their Effect on Beaches

His adrenaline surged a bit as Chuck caught his first wave of the morning. He turned right along the wave face and got a 15-second ride with the sunrise just beginning to peek over the houses. What a way to start the day! Within two hours Chuck would be in his car creeping up Interstate 5 on his way to the office. But now he was part of the "dawn patrol," the thousands of men and women with "real jobs" that surf in the early mornings in Southern California. And for just a few seconds, his entire focus in life was on staying on that wave as he accelerated to over 25 miles per hour. Chuck would quit surfing if he ever stopped feeling that little surge of adrenaline.

A wave breaks along every one of America's beaches every few seconds. Some are small waves that lap gently against the shore. Some are huge monsters that crash violently with the power to move practically anything. Surfers will catch a few of them. People will watch some of them from the shore. But all of them will break and expend their energy as they race up the beach. And then there will be another and another and another forever.

Waves are the primary force shaping our beaches and coasts today. Anyone who has ever had their breath knocked out of them by a wave in the surf has a visceral understanding of the force in waves. The power and energy in those waves can be mind boggling during storms. Bigger waves are more powerful. The size of waves can be measured by their height and length. The oceanographer's definition of wave height is the difference in elevation between the lowest part of the wave, the trough, and the highest part of the wave, the crest. Wave energy is proportional to the wave height squared. Thus, a 6-foot wave has four times the energy of a 3 foot wave. And some waves heights can exceed one hundred feet!

Riding waves on a raft is often a very easy way to enjoy the surf.

The scientific study of waves changed during World War II. Plans for amphibious landings such as at Normandy on D-Day and on the Pacific islands required as good a prediction as possible of the surf conditions that the landing craft could expect. Then-secret research efforts led to equations that forecast wave heights based on wind speed. The research also looked at how waves change as they move into shallow water and break. This research was released to the public after the war and revolutionized nearshore oceanography. Today, surf forecasting benefits surfers, boaters, lifeguards,

The splash from this wave is over 60 feet high. The wave is hitting a rock stack off the California coast. During storms, waves have inundated lighthouses over 100 feet above sea level.

beach managers and coastal engineers.

Almost all the waves that break on beaches were previously generated by winds. A few were generated as boat wakes and a very few were generated by underwater landslides. Wind first ripples the water surface and then begins to make waves. Wave heights continue to increase as the wind blows farther across the water. Stronger winds make higher waves. If the water body, the bay or ocean, is large enough, eventually the wave heights will stop growing unless the wind speed increases.

There is another way to measure size of waves. Wavelength is the distance between two crests in a train of waves. Wind ripples have very small wavelengths, sometimes as short as a few inches. But as waves get bigger in height, they also get longer in length. Wavelengths can grow to over a thousand feet.

The wave "period" is the time between waves. For example, if you are watching waves with a 10-second period, one wave will move past the end of a fishing pier or break on the beach once every 10 seconds. It turns out the longer wave periods occur with longer wavelengths.

Windsurfing in large surf combines two extremely challenging sports and makes for great entertainment for spectators. The breakers in this picture are spilling breakers from locally generated "sea" waves. The wind was onshore (blowing in the same direction that the wave was moving) at about 20-25 knots.

The motion in an ocean wave is a thing of beauty and endless fascination. The only thing that is moving very far is the waveform itself. The actual water particles just move in circles while the wave, the disturbance of the water surface, moves past. A float or a small boat will just ride forward and up and then backward and down as a wave passes under it. This is like sound waves that move great distances through the air while the individual air molecules just move, actually vibrate, a little. It is also like the "waves" in a blanket or sheet that is being shaken across the surface of a bed, only the disturbance or ripple in the blanket really moves very far. As water waves move into shallower water, the circular movement of the water changes to an elliptical shape. Near the bottom in shallow water, the water just moves back and forth as each wave passes over. This is called "surge" by divers and can be seen by watching grasses near the ocean floor sway back and forth.

Waves often travel for hundreds or thousands of miles across the ocean once they are generated. This means they travel outside of the storm that generated them. Most waves that hit the beaches of America were generated far out at sea in either the Atlantic or Pacific. The storm that created the waves is often not even visible where the wave breaks on the beach. These waves, ones that have traveled out of the winds that generated them, are called "swell" waves. Waves that are still being acted on by the

Hurricane Georges destroyed this condominium on the Gulf of Mexico in 1998.

winds that created them are called "sea" waves. Both types of waves, "swell" and "sea" can be extremely powerful. Hurricanes as well as the El Niño storms of the Pacific and the northeasters of the Atlantic bring unusually large sea waves with their winds and also send large swell waves across the ocean.

There are a few types of waves that are not generated by winds. Boat wakes are obviously created by the disturbance of the moving boat. The wake off the bow, or front, of large ships can be over 10 feet high. Dolphins sometimes play in these waves. They "surf" the wakes and sometimes make tremendous leaps out of the sea.

Tsunamis are waves generated by underwater landslides during earthquakes. The problem with these waves is that they have extremely long wavelengths that allow them to break up onto land much higher than normal wind-generated waves. A tsunami that hit the Pacific coast of America in 1964 killed 12 people and caused millions of dollars in damages in northern California and Oregon. In 1946, a tsunami slammed Hilo, Hawaii killing 150 people. All of America's coasts are subject to the possibility of tsunami waves that can cause catastrophic death and destruction. Tsunamis are sometimes improperly called "tidal waves" even though they have nothing to do with the tides. The Japanese word for the phenomenon, "tsunami," was originally borrowed by oceanographers in order to avoid confusion with the tides. There really is, however, a "wave" related to the tides.

Tides

The tide is the slow rise and fall of the ocean waters in response to the gravitational pull of the moon and the sun. The tide is a type of wave! The tide is a very, very long wave with a wave period of 12.4 hours. The tide "wave" travels around the ocean basins in a circular motion like a sloshing bathtub. The arrival of the crests of this wave represent high tide and the troughs represent low tide. Since the moon's influence is about twice as strong as the sun's influence, the tide mostly follows the lunar calendar. But the highest tides occur when the sun and the moon are lined up and the two forces work together. Full moons and new moons bring the highest tides, sometimes called spring tides, for this reason.

There are usually two high tides and two low tides every lunar day (24.8 hours). These are called "semi-diurnal" tides with about 6.2 hours between high and low tide and 12.4 hours between successive high tides. The time of the tides slips about 48 minutes every day. So, a high tide at 10:00 a.m. on Monday would mean that high tide would be about 10:48 a.m. on Tuesday.

Large differences in tides occur around the country because the tide "wave" runs around each ocean basin and is warped by the shape of the continent differently. The tide range, the elevation change between high and low tide, differs tremendously along the coast. The tide range in Calais, Maine is about 20 feet! This means that the water level drops 20 feet in a little over 6 hours and then rises again in the next 6 hours. In some parts of Alaska like Anchorage, the tide range is even larger, up to nearly 30 feet. The tide range at Sandy Hook, New Jersey is about 5 feet but it is only 2 feet at Montauk Point, New York just 125 miles away. In general, the higher tide ranges occur in the northern latitudes and the lower tide ranges occur on islands that are away from the continent.

One interesting difference in tides is that on the Pacific coast the two high tides that occur during the day usually have very different elevations. One is much "higher" than the other high tide. Oceanographers call this a "diurnal inequality." Both high tides are usually about the same elevation on the Atlantic coast.

The Gulf of Mexico has some very unique tides. There are usually only one high tide and one low tide each day. This "diurnal" tide also has a very low range of about 2 feet or less. The strange tides in the Gulf of Mexico occur because of its shape and because it is almost an enclosed sea. The only connections with the Atlantic Ocean are the narrow straits on both sides of Cuba.

The Great Lakes are tideless. They are completely enclosed and isolated from the Atlantic Ocean tides and they are too small for any lunar tides of their own. Water levels in the Great Lakes fluctuate very slowly in response to rainfall in that part of the continent. There is an annual rise and fall of between 1 to 2 feet on Lake Erie due to snowmelt runoff in the spring. However, during multi-year dry or wet periods that seem to come each decade, the lake level rises and falls in a range that varies around 3 to 5 feet. During the last century, the lowest monthly average lake levels were in 1934 and the highest were in 1985. The difference between those two extremes was over 5 feet.

During great storms, the lake levels can rise or fall up to 10 feet! Wind pushes the water from one end of the lake to the other.   A storm in 1985 pushed the water in Lake Erie so well that the water level dropped eight feet in twelve hours at Toledo, Ohio at the western end of the lake. Meanwhile, at Buffalo, New York at the eastern end of the lake, the water level rose almost eight feet.   In the middle of the lake, at Cleveland and Fairport, Ohio, the water level only fluctuated a little. "Seiching" is a phenomenon on the Great Lakes that happens after a storm like that one passes across the lake. The water sloshes back and forth in the lake after the storm like water sloshing around in a bathtub.

Storm waves

Old-timers in east Texas used to know when a hurricane was coming across the Gulf of Mexico because the swell waves from the storm moved faster than the storm itself and was a crude warning system. This "swell forerunner" would send unusually

long period waves to the beaches about a day before the winds from the hurricane.   The swell forerunner the day before the killer Hurricane of 1900 brought very large, very long period waves to the beaches of Galveston.   Wave periods, the time between waves, were probably about 12 seconds.   Most Galveston waves have periods of only about 5 seconds so the swell forerunner was most unusual in every way: height, period, and length.

The Galveston Hurricane of 1900 was the worst natural disaster in American history. Between 6,000 and 12,000 people died when the storm surge allowed the hurricane waves to ride up across the low barrier island. (From *Story of the 1900 Galveston Hurricane* by Nathan C. Green, © 1999 used by permission of the publisher, Pelican Publishing Company, Inc.)

It wasn't so much the size of the waves, although they were huge for the Gulf, but the storm surge that raised the water level upon which the waves rolled across the island that was the killer. Before the sun rose the next morning, between 6,000 and 12,000 people were dead in our nation's worst natural disaster. Buildings were pounded down by waves and then the people drowned in water depths that approached ten feet in parts of town that were normally five feet above high tide. In the decades after that killer hurricane, Galveston built a large seawall and raised the elevation of the island. Subsequent hurricanes have not been nearly as deadly. The seawall has been successful in protecting lives and buildings. However, the seawall has not done anything for the beaches of the city.

Storm surge is the combination of all the ocean's phenomena that make the water level rise above its normal tidal elevation. It includes the effect of low atmospheric pressure systems. It includes the water that is literally blown across the ocean by wind stress. It includes some additional elevation due to waves breaking in the surf called wave setup. And it occasionally includes rainfall and runoff in the coastal areas. But storm surge is not a "wall of water" as it is often called. It is a fairly gradual rise in water level that can last for 4 or 5 days. However, killer hurricane storm surges often have an extreme rise in water level over a 10 to 30 minute period that then goes back down almost as quickly. Six to ten foot storm surges are fairly common during major hurricanes in America. A few have been measured as high as twenty feet. The storm waves then ride across that raised water level causing damage.

Surfing is a balancing act!

During the storm that Sebastian Junger named "the perfect storm" in his modern epic, waves were measured at a gage in the Atlantic that had individual wave heights exceeding 50 feet. There were certainly much higher waves that were not measured. The storm was legendary in the coastal engineering community and the surfing community long before the publication of Junger's book. It was called "The Halloween Storm" since it hit on Halloween of 1991. Swell waves from that storm hit the beaches of south Florida at heights that no one alive had ever seen there. The pictures look more like the north shore of Hawaii than the Atlantic Coast! The storm caused beach erosion and inlet movement and creation at many places along the eastern seaboard. It also damaged piers, jetties and beach nourishment projects. Even more extensive damage was caused by a northeaster storm that hit on "Ash Wednesday" of 1962. That storm lasted three days and included five high tides. The damage was tremendous along all of the mid-Atlantic coast barrier islands.

The El Niňo storms of the Pacific coast cause similar damage. The winter of 1982-83 brought unprecedented storms and waves to the Pacific coast. We now know that these great storms were part of the ENSO, for El Niňo - Southern Oscillation, phenomenon related to the interactions between the oceans and the atmospheric circulation patterns. At the time, however, coastal Californians just knew that they were getting pounded. A series of six major storms came ashore in three months. They brought some of the greatest rainfall, the strongest winds, and biggest waves on the highest ocean water levels of the century. Sustained winds exceeded 60, 70, and even 80 miles per hour for much of the coast in several of the storms. A gust of 96-mph was measured at Pillar Point in California during one storm. Individual wave heights exceeded 40 feet and the average wave height at Monterey California was measured above 15 feet for each storm. The storms caused damage to piers, houses, seawalls and beaches.

The surfing culture includes some different definitions of wave height. On the huge waves of the north shore of Hawaii, the local surfers estimate size by looking at the back of the wave as it breaks. This number can be only about half that of the oceanographers' definition. Thus a surfer's 10-foot wave is really a 20-foot breaker as defined by oceanographers! Surfers are notorious for underestimating wave heights. Another common way for surfers to talk about wave height is in relation to the top of the wave as they catch it, such as "waist high" or "head high." A huge surf might be

This New Jersey lifeguard is riding waves in his surf rescue boat. He initially rows to catch the wave and then ships one oar and jumps to the stern to steer using the other oar like a rudder. Even on this small wave, this ride was over 100 yards long. When the waves are larger, these surfboats can ride over a half-mile in to shore. If they aren't steered absolutely straight down the wave, they will broach sideways and flip over! This makes for some exciting moments in lifeguard races in big surf.

"double overhead" or twice as tall as the surfer, i.e. 10-12 feet high. People watching world-class surfers on television don't realize what kind of athletes these men and women are because they make it look easy! Consider surfing on a 12-foot high wave. When surfers first paddle to catch such a wave, they stand up near the top and look down about 17 feet. For a similar perspective, stand on the edge of the roof of your house and imagine the balance needed to slide down the side of the house. If that's not enough to scare you, now image that your house is moving at 30 miles an hour as you do it and will land on you if you wipe out!

Thousands of American surfers have learned how to surf on swell. Southern California, the hotbed of surfing in America, has swell waves from the south hitting their beaches every summer. These waves were generated by large winter (winter in the southern hemisphere) storms in the southern Pacific. These swell propagate north several thousand miles to the beaches of San Diego and Los Angeles where they break in perfect tubes when the local winds are offshore. The famous waves of the north shore of Hawaii are formed in the fierce winter storms of the Bering Sea. They move south across the

Rocks covering an old, submerged pipeline that runs offshore are causing these waves to peak up in Santa Monica Bay. Surfers ride this "break" and many others formed by man-made structures including groins, jetties, and piers. Several experimental surf reefs have been built to enhance wave breaks for surfing.

northern Pacific Ocean before hitting Wiamea Bay. Depending on the severity of the storms in the Bering Sea, the waves vary from almost non-existent in the summer to over 50 feet high during "big wave" contests. One contest is held only when the waves exceed 50 feet (the surfers call it 25 feet of course). That contest is not held every year because those conditions don't occur every year.

The story of Mavericks is a modern legend. Some of the biggest waves ever seen hitting America's coast have been seen there recently! Some of the biggest waves ever surfed have been surfed there. Mavericks is the name of a surf break offshore of Pillar Point on the northern California coast south of San Francisco. Until a few years ago, no one had ever even heard of the place! In spite of its proximity to the metropolitan San Francisco area, it is fairly remote. Access to Pillar Point is severely restricted because it is a military installation. Also, on most days, there are no waves breaking there.

However when the wave conditions are just right the swells begin to break across a deeply submerged rock reef. Some of the local surfers used to sneak out to Pillar Point to watch these "mountains of water" break and dream about someday taking a boat out there to ride them. The first one to do so was a local guy that also surfed the big waves in Hawaii. And for a while he had Mavericks all to himself! Apparently, he invited some of his buddies on the professional big-wave surfing tour and the rest is history. At one televised contest at Mavericks, one of the top surfers in the world tragically drowned after wiping out. Now, many rides on these waves begin with the surfer getting towed by a jet ski to catch the wave. In November 2001, several waves with faces as big as 70-feet high were surfed. Later that same day several people watched a set of waves estimated at 100-feet high. These may have been the largest waves ever to hit the American coast. But more likely, they are the largest waves ever to be seen: there are probably larger ones that weren't seen because no one was there to see them!

These wave crests are bending as they move into a pocket beach in Andrew Molera State Park in California. This bending is due to both refraction and diffraction.

Breaking waves

Between the deep ocean and the beach, waves undergo some dramatic changes. Once a wave reaches a depth that is about equal to half its wavelength, it begins to "feel" the bottom. It begins to change its speed, size, shape and direction. These changes continue until it breaks in shallow water and then runs up the beach. The first thing that happens to a wave moving into shallow water is that it slows down as it feels the bottom. Wave speeds that are commonly 30 miles per hour and can exceed 50 miles per hour in deep water typically slow down to about 10 mph at breaking and 5 mph in ankle deep water. When a wave approaches the coast at an angle, part of it is in shallower water than another part of the same wave crest. The wave crest will appear to be bending since the shallower part is slowing down more. The term that is borrowed from optics for this bending is refraction. In optics, refraction is the bending of light waves by eyeglasses and corneas. For water waves, refraction is the bending of the wave crest as it moves into different water depths. A closely related change is

The waves in the foreground are breaking across a submerged, hard rock outcropping called Table Top Reef in southern California. In the background is a sandy beach across the mouth of a coastal lagoon.

diffraction or spreading out of wave energy. Diffraction explains why wave crests spread out as they enter a cove or harbor. The wave crest wraps around into the sheltered areas.

As waves bend, they change direction. The wave energy can be focused onto a headland or defocused from a cove. The spectacular breakers on Mavericks and other famous surfing reefs are the result of refraction focusing wave energy. On the other hand, several ports on the Pacific coast have been built to take advantage of areas where ocean waves just can't get usually. Point Arena, California has a pier that sticks out into the Pacific Ocean. The pier is in a cove that defocuses most of the wave energy. The pier is also in the lee of several offshore rock reefs that cause the waves to break reducing their heights even more. It is not uncommon for the waves at the pier to be less than one foot high when the offshore waves are over 10 feet high. The Point Arena Harbor Pier has no man-made breakwater protecting it. It works during most wave conditions but in unusually high wave conditions, the port cannot be used. During the El Niño storm of January 26, 1982 a set of three huge waves crashed over the pier. Large waves usually move in "sets" or "wave groups." The sets have 3, 5, or 7 large waves moving together.

Because of its geology, the Pacific coast of America does not have many natural, enclosed bays for harbors like San Francisco and San Diego. Thus, many piers were used as ports over the years. Some of them, like the pier in Newport Beach, California, were built as shipping wharves but have been turned into pedestrian piers as the shipping industry went elsewhere. Some ports, like San Pedro (Long Beach) and Santa Barbara added offshore rock breakwaters to provide an even better harbor.

If a wave moves for long distances across shallow water of constant depth, it will lose some energy and height. This dampening across mudflats and refraction is one of the reasons the "armpit" of Florida, the Gulf of Mexico coast between Tampa and Apalachicola, doesn't have sandy beaches. Almost all the wave energy is lost and the waves are so small that mud banks can form along the shore. Sandy beaches require

some energy to remove the silts from the beach sediments.

Waves break for two related reasons. In deep-water, when the wave height grows until it equals about one-seventh the wavelength, the waveform becomes unstable and the wave breaks. The breaking is often seen as white-capping when the wind is still actively blowing on the sea. In shallow water, waves break when

This small wave is plunging. Note how the overhanging jet touches down in front of the wave causing the white-water splash. On much larger waves, this jet has so much force it has been known to snap unfortunate surfer's boards or bones!

they reach a depth that is almost equal to the wave height. Near that depth, the waveform becomes unstable and the wave breaks. Thus, breaking waves near the shore show where the water first becomes very shallow.

There are two common types of shapes that waves take when they break on beaches – spilling and plunging. The spilling breaker begins gradually at the top of the wave crest with some tumbling white-water. The tumbling water increases until the entire wave face is tumbling white-water. A plunging breaker stands up in a vertical wall of "green water" and then the crest continues to plunge forward until it lands down lower on the wave face. This plunging creates a "tube" of air for a brief second. This is the tube that surfers try to ride in. The plunging breaker is more violent than the spilling breaker and stirs the water up more. The turbulence associated with plunging waves often extends down through the water to the bottom and can pick up a trench of sand and throw it into suspension. Close inspection, including time-lapse photography, of spilling breakers reveals that they actually start with a very small plunging portion. The jet is so small and quickly broken that it is difficult to see. Breaker type is controlled by three

The Coast Guard trains in the big swells near the mouth of the Columbia River.

things: wave steepness, beach slope, and local wind direction. Wave steepness is the ratio of wave height to wave length. Swell waves have lower steepness and tend to plunge more. Sea waves tend to spill more because they are steeper. The steeper the beach slope, the more the breakers tend to be plunging breakers. The wind at the beach is the final piece of the breaker type puzzle. Onshore winds, those blowing from the sea toward the shore, tend to cause breakers to spill. Offshore winds tend to cause breakers to plunge. Since most surfers prefer plunging waves because they like to slide across the vertical face, they prefer swell, steep beaches, and offshore winds. One of the original oceanography textbooks even jokes in a footnote about the absence of graduate students at the University of Hawaii from class when the wind is offshore. They are all out surfing! Or excuse me, studying surf zone dynamics!

There are two other breaker types that rarely occur on sandy beaches. Surging breakers are waves that reflect off vertical walls. Collapsing breakers stand up and then the entire wave face moves forward up a steep slope. They are the most damaging breaker type to some coastal structures and they can occur on very steep beach slopes. When large waves are collapsing on a beach it is impossible to stand in the surf.

Because of the size of the ocean, long period swell is more common on the Pacific coast of the United States. The Pacific Ocean is larger than the Atlantic and so there is a higher chance that some wave energy will be propagating across it from some distant storm. Wave periods of 8 to 16 seconds are common and very long period swell can be as long as 24 seconds. On the Atlantic Coast of the United States, common wave periods range from 6 to 12 seconds. In the Gulf of Mexico, wave periods are typically 4 to 8 seconds; swell from distant storms doesn't even always exist. Locally generated seas are often the dominant waves. Because the Great Lakes are so small, swell rarely is significant and locally generated seas are almost always the dominant waves.

Just before a wave breaks in shallow water, it rises up and changes shape. The increase in wave height is called shoaling. Its overall shape changes as the wave crest gets more peaked and the wave trough flattens out. The combination of these two effects explains why fishing piers often get damaged about halfway out the pier instead of at the end of the pier. Large waves shoal up and destroy the horizontal decking of piers where they approach their nearshore breaking depth. At the outer end of the pier, the waves haven't shoaled up as high. Waves can shoal up to over 150% of their offshore height.

Some of the biggest waves in the United States break on the Columbia River bar. The breakers are so big, so consistently, that the US Coast Guard has a base there for training. The sand bar outside the mouth of the Columbia River is the ebb-tidal shoal. Sands that come out of the Columbia River and off the adjacent beaches of Washington from the north and Oregon from the south get pulled out there by the river currents and the ebb-tidal currents. The combination of the shallow shoals and the outgoing currents combine to make the waves stand up and break consistently. Ocean currents alone can make waves change size and shape. If an ocean current is moving in the opposite direction of the wave, the wave height increases. This combines with the shoaling of the large waves typical of the northern Pacific at the Columbia River bar. These breakers can dwarf the Coast Guard cutters. These amazing ships are meant to ride over and through these mountains of water and to be self-righting if they do get rolled over in the surf.

The accelerations in breaking waves can exceed two or three times the acceleration and force of gravity. The highest accelerations occur low down on the face of plunging breakers. This is exactly where dolphins ride waves in the surf. Since the dolphins are not feeding, it appears that they are playing when they ride waves. They must get a kick out of feeling these high accelerations. You could say this is Mother Nature's original roller coaster ride!

Most wave riding is a balance between this acceleration and the rider. Body surfers, boogie-boarders and surfers use the high acceleration zone to "take-off" and then they turn along the face of the wave to extend their ride. Once a rider obtains enough forward speed, the board begins to plane across the surface of the water. Body surfers usually can't plane as well on their stomachs as with a board and their rides are usually shorter.

Rip currents

Rip currents are killers. As waves break in the surf, some of their energy gets transformed into currents. These currents are different from and in addition to the wave motion surging back and forth. The most famous of these surf currents are the rip currents and the longshore currents.

While statistics concerning beach drownings are hard to come by, they are clearly a significant problem. They occur along all of the Pacific, Atlantic and Gulf coastal states. According to research at the University of Florida, there are an average of 21 beach drownings per year in Florida.

Many of the beach drownings in America are caused by rip currents. Rips are currents that flow away from the beach. Rips can take a person from a water depth where they can comfortably stand up out to a depth where they can't. Rip currents can flow at over 5 miles per hour and this is faster than even an outstanding swimmer can maintain for long. Fortunately, rip currents are fairly narrow and don't go very far anyway. They are usually only about 20 to 60 feet wide and usually only flow about 100-500 feet offshore. Some California rips have been reported to extend up to one-half mile offshore.

People drown in rips because they panic. They panic because they can't swim at all or they try to swim against the current and still get pulled offshore. The most common advice for people caught in rip currents is to swim parallel to the beach. This will get you out of the current, and if you can't stand up on a sand bar there, you can then swim in toward the beach successfully.

Another piece of advice is to just "enjoy the ride!" Just don't panic. If you're ever caught in a rip current, pick up your feet, tread water, laugh to yourself and enjoy the ride. Laugh because being overweight is saving your life! Most Americans float pretty well in salt water because we are so fat! Salt water increases the buoyancy of the human body. Besides, the rip current won't take you very far, rarely more than 100 yards (and if

There was a small rip current through the shallow sand bar when this picture was taken. The rip current is in the center of the picture and flowing offshore towards the bottom of the picture. You can see how it is making the waves break farther offshore.

you can't swim that far you shouldn't be in the surf at all), then it will stop outside the breaker line. Then you can drift or swim down the beach and then swim into the beach where there is no rip current. On guarded beaches there will be lifeguards on their way to you before you get to shore. Wait for them and don't be embarrassed about being rescued. Don't tell them that you enjoyed the ride - they might not appreciate that!

Rip currents often occur where a sand bar has a low spot. Breaking waves move water over the sand bar into the deeper water on the inside of the bar. This deeper water is called a trough or gully. The water in the trough flows down the beach and back out to sea through the low spot in the sand bar. The location of the rip current is sometimes visible from the beach as a place where the waves are not breaking the same as elsewhere on the bar. They might not be breaking at all or they might be breaking farther offshore and with a different breaker type because of the outgoing current. Sometimes, a cloud of suspended sand in the rip current is visible. This sort of rip current stays in one place for hours or days or even months in a row. Swimmers on the sand bar or drifting down the beach can drift into one of these rip currents. Since rip current energy is provided by the breaking waves, they don't always exist. If they do exist, the strength of the current will change throughout the day as bigger waves come and go.

If you watch an experienced swimmer or surfer approach the water, you will notice that they almost always study the surf for a few minutes or longer before entering. Some things that they are judging include how big the waves are, how they are breaking, whether or not there are rip currents, and where to go in to either avoid them or get in them. Sometimes surfers will use rip currents to get through the breaker line with the least expenditure of energy paddling out.

Another, more dangerous, kind of rip current can occur at any time and any place when there are large waves coming ashore. Water that was not flowing can begin to flow and pull a swimmer, who was fine a minute ago, offshore rapidly. These rip currents, called "temporal rips," turn off and on and sometimes drift down the beach. These usually occur when large waves are coming directly into the shore. Many lifeguard rescues are needed on days when this type of rip current is happening. Hundreds of rescues have been reported on the same day.

Note that rip currents are not directly related to the tides. A rip current's energy is provided by the breaking waves. Thus the term "rip tide" is a terrible term. It's not an "undertow" either. It's not towing you under but just offshore into deeper water. Call it a rip current and remember, if you ever think that you might be caught in one, laugh and enjoy the ride.

The offshore sand bar is causing the waves to break twice in the above picture of Daytona Beach. The whitewater on the right of the photo is the breaking on the 4-foot deep sand bar. Then the waves cease breaking as they move across the 7-foot deep trough or gully and then break again on the shore.

A "river of sand" — longshore sand movement

The other famous current in the surf is the longshore current. This current is famous because of its ability to move sand down the coast and change shorelines. Longshore currents are fairly constant currents that flow along the beach parallel to the shoreline inside the surf. Longshore currents occur because waves break at an angle to the shoreline. Some of the energy in the breaking waves drives the current. Longshore currents explain why swimmers drift down the beach. The speed of the longshore currents can approach 5 miles per hour but is usually less. The longshore current only extends out across the surf zone and stops beyond the depths where waves are breaking. There are sometimes much weaker currents moving along the beach out beyond the surf zone that are due to wind stress or oceanic tidal movements. Longshore currents aren't particularly dangerous to swimmers except when they move the swimmer down the beach into an existing rip current.

Longshore current is not always in the same direction. Since it is driven by the incoming wave angle, it often changes directions. For an observer standing on the beach, if the waves are approaching the beach from the left, the longshore current will be to the right. If the waves are approaching the beach from the right, the longshore current will be to left.

The most important aspect of the longshore current is that it can move tremendous amounts of sand down the beach. The incoming waves agitate the sand on the bottom of the sea and breaking waves kick some of the sand up into the water. Then while the sand

Waves move tons of sand down the beach in a "river of sand." Note how the waves are approaching the Atlantic City beach at a slight angle in this 1985 photo. On this day, the incoming waves were driving both the longshore current and the longshore sand transport to the south or to the left.

drifts back down, the longshore current moves it down the beach. Thus the longshore current acts like a conveyor belt and the waves throw the sand up into the conveyor belt. This sand movement is called "littoral drift" or "longshore sand transport." The process has been likened to a "river of sand" that moves down the coast. The banks of the river are the breaker line, since wave breaking drives the current, and the beach. This "river of sand," however, flows in both directions, left and right along the coast, depending on the direction of wave approach. When waves come from the left, sand moves to the right. When waves come from the right, sand moves to the left. And when there are no waves, there is no sand movement. The amount of sand moved can be tremendous. During big storms the rate can exceed 100,000 cubic yards per day moving down the coast. This would be the same as a line of bulldozers end-to-end slowly pushing sand down the beach all day and all night. Most of the sand moves during storms. In some places, 90% of the sand movement occurs during only 10% of the time.

The sand bar is often completely exposed at low tide. The birds on this sand bar in Homer, Alaska are bald eagles.

Most beaches in America have longshore sand transport in both directions. Usually, one direction is dominant but at many places, the transport is fairly balanced. In other words, almost as much sand moves to the left as to the right. Longshore sand transport is an important geological force. It is responsible for baymouth bars sealing off river mouths in the Pacific and in New England. It is a dominant force shaping the barrier islands of the Atlantic and Gulf too. Sand transport over years and thousands of years can reshape and move the barrier islands. Sand moves down the beach until it gets to an inlet. Then it often gets moved out onto the ebb-tidal shoal bar. The sand in the ebb-tidal shoal then bypasses the inlet in the form of sand bars migrating onshore onto the downdrift island.

If longshore sand transport gets interrupted for any natural or man-made reason it can cause erosion for miles downdrift. This is one reason that works of man have caused so much beach erosion. Man-made interruptions of this "river of sand" have led to most of the severe erosion problems in American history.

Sand bar movement

Besides moving down the beach, sand also moves offshore and onshore. The underwater sand bars that are common on many sandy beaches are evidence of this. These sand bars often move onshore during periods of low wave heights. Research has shown that it is actually wave steepness, the ratio of wave height to wavelength, that controls sand bar movement. When the wave steepness is low, such as with swell waves, the sand bars typically migrate to the shore. The sand bars sometimes move all the way into the dry beach and make the dry portion of the beach wider. This usually happens when there is a long period, long wavelength swell with small wave heights hitting the beach. These conditions often occur during summer along most of the American coast. For this reason, a wide beach is often referred to as a "summer" beach profile. When waves are steep, such as with a short-period wind sea, sand is pulled off the dry beach onto a sand bar. Often, existing sand bars are pulled even farther offshore during these conditions. These sea wave conditions often occur along the American coast in winter storms. Thus, a narrow dry beach with a well-defined offshore sand bar is typically called a "winter profile." The shift from a wide, summer profile to a narrow, winter profile often occurs during the first one or two storms. After the "winter" profile is established, subsequent storms usually don't move more sand unless they are more severe. The beach profile reaches "equilibrium" with the incoming waves. However, since waves change constantly, the beach profile rarely gets to any real equilibrium but is constantly adjusting to the changing incoming waves.

This seasonal shift in the beach really has nothing to do with the temperature but with the waves. And even though it appears to be causing beach erosion, it is really just a natural, normal shift in the beach that doesn't cause any real long-term shoreline change. People remember the wide beach that they sat on in July when they come back to the beach in November and think that most of it is "gone." Most beach erosion stories in the press run during the fall or winter when the dry, visible beach is at its narrowest naturally. These seasonal fluctuations in dry beach width caused by cross-shore movement of sand are not the cause of real beach erosion. Cutting off the movement of sand along the beach causes most real beach erosion. If longshore sand movement is interrupted, beaches can erode for miles.

Small beach cusps, the repeating patterns in the wet sand here, were formed on this beach during the previous high tide. Scientists really aren't sure why these occur!

Chapter 4

"Sand Thieves" of the Beach —
How We are Destroying Our Beaches

> Travis and Paige are getting desperate. For years, each storm has eaten a little more of the sandy beach away and the beach house they bought for retirement 20 years ago is now severely threatened by erosion. They don't want to build a seawall to protect the house but they can't move the house back either. Some folks tell them that all beaches erode and that they shouldn't try to live there in the first place. But they suspect the real cause of the erosion is the shipping and boating channel just down the coast. The channel, which has jetties and is dredged regularly, provides safe access to the sea for boats. While Paige and Travis have nothing against boating and shipping, they wonder why they have to lose the sandy beach in front of their home and maybe even their home for the benefit of those that use the channel. It's just not right.

The primary cause of much of America's beach erosion is sand robbery! Sand often gets removed from the beach when it gets trapped in another part of the beach system. The result is the same as taking it directly off the beach – beach erosion. The sand theft is not obvious or illegal but rather unseen and often unrecognized. It is occurring somewhere else and usually very gradually.

The sand thieves are "works of man" including dredging of ship channels without sand bypassing, trapping of sand with jetties, groins, seawalls and behind dams on the west coast, and even sand mining. All of these remove sand from some part of the beach system. All of these "works of man" are being done for some other good reason. But where the problems are not being addressed, they are slowly and quietly stealing the lifeblood of America's beaches – the sand.

The amount of sand removed from the beaches has apparently never been totaled up before. While it is well known that many individual "works of man" have removed sand from nearby beaches no one has ever added up the volume of all of the nation's individual beach sand thefts. Doing so leads to the following stunning conclusion.

Over a billion cubic yards of sand have been removed from the beaches of America by "works of man!" This is a tremendous amount of beach sand. This much sand is enough to fill a football field over 100 miles high. If all the sand removed by "works of man" sand were restored and spread evenly along 2000 miles of coast, a distance roughly equal to America's developed sandy beaches, the beaches would be about 70 feet wider.

The Port Canaveral harbor, artificially built in 1948, caused beach erosion for miles to the south. Not only did the jetties interrupt the movement of sand to the south beaches but sands dredged from the channel were thrown away offshore for decades. Within the last several years, the beaches to the south, Cocoa Beach, have been nourished and sand has been bypassed around the jetties.

Man-induced sand removals from the beach system are a major cause of beach erosion in America. In many locations, they are <u>the</u> major cause of beach erosion. The impact of sand robbery varies throughout the country but the most severe, and most publicized, individual erosion problems in almost every state are due to some form of sand robbery. Shorelines fluctuate in response to changes in waves and water levels too and the natural fluctuations often mask the man-made signature. Also, estimates of the amounts of sand removed from the beach system are very uncertain. But the magnitude of the robbery probably exceeds that of natural forces in the past century.

The beach sand stealing continues to this day! Many of the worst cases have been stopped with sand bypassing attempts and some have even been reversed through beach nourishment. But millions of cubic yards of sand per year are still being removed from the littoral system throughout America. Some of these removals cannot be stopped. For example, the reductions in sand moving to the southern California beaches from the coastal rivers that have been dammed and lined probably cannot be avoided. However, some of the sand removals can be stopped. We are still unnecessarily dumping beach sands dredged from some ship channels offshore or piling it up onshore. The future health of America's beaches requires that this be stopped.

We need laws developed and applied that restore dredged beach sands to the beach where they can continue to move down the coast in the littoral "river of sand." It is

the "sustainable" way to maintain channels and beaches. Florida has such a law and it has been effective. The beach sands of America should be treated like a valuable resource. They are. They are the essence of the beach. We are destroying our own beaches.

Sand "budgets" are a good way to think about beach erosion. Sand along our beaches is like money in a savings account. There is only a limited amount of sand and if some of it is removed the beaches

Much of the beach erosion in America is due to ship channels that interrupt the movement of sand along the coast. The beaches south of the entrance to Port Everglades, Florida are chronically eroded.

will suffer. If the withdrawals are larger than the deposits, a given stretch of beach will erode. The sand budget idea is typically applied to a given stretch of beach. Widening beaches have more sand entering the area than leaving the area. Eroding beaches have more sand leaving the area than entering the area. This chapter outlines some of the man-induced withdrawals from the sand budget of America's beaches.

Ship channels as sand thieves

Ship channels rob sand from beaches several different ways. One of the most obvious is offshore disposal of dredged sands. Dredging is required for safe navigation for pleasure and for commerce at almost every American port. The problem is that some of the shoaling sediments that must be dredged are beach sands that were moving in the wave-driven "rivers of sand" along our coasts. Along the coast, most ship channels must be dredged through a sand bar that is part of the beach's littoral, sand-sharing system.

Port Canaveral, Florida is one example of how ship channels can rob sand from beaches if the dredged sand is not placed back on the beaches, i.e. bypassed. Over 8 million cubic yards of sand have been dredged from the channel and dumped offshore in deepwater away from the beaches. The port was built in the 1950's to stimulate the regional economy but in recent years it has become one of the western hemisphere's busiest cruise ports. It has been the start of millions of cruise vacations! There was no natural inlet through this beach so the engineers created a channel by dredging through the beach and dredging a port behind the beach. Rock jetties were built to hold the channel in place.

Even with the jetties, sand moves from the north beaches and, to a lesser degree, from the south beaches, into the channel. So about 200,000 cubic yards of sand and silt must be dredged each year to maintain the depths needed for the ships. For decades, that sand was hauled several miles offshore and dumped. Considering the sand budget for the area, the total sand withdrawal is over 12 million cubic yards. This includes sand trapped in the build-up of a wider sandy beach on the north side of the jetties. The downdrift beaches, including Cocoa Beach and Satellite Beach, eroded because of this port. It essentially cut off their sand. The "river of sand" along this coast was not getting to those beaches and they were suffering. The Canaveral Port Authority recognized the problem years ago. They have been working with the downdrift towns to fund beach nourishment projects to restore the damaged beaches and working to bypass the sand around the channel to the downdrift beaches in the future.

This artificial channel cut in 1917 has caused beach erosion downdrift for miles. Sand moves to the south along this coast. The beaches on the north side of the inlet jetties have gotten much wider while the beaches on the south side have eroded. This is Lake Worth Inlet, Florida (circa 1965).

Bob Dean and his colleagues at the University of Florida pointed out several decades ago that this problem was damaging the beaches of Florida. They estimate that over 50 million cubic yards of sand have been dumped offshore from the inlets on the east coast of Florida. This explains about 85% of the beach erosion losses along that coast! There is little offshore dumping of sand in south Florida today because of the state law that prohibits it and requires instead that sand dredged from channels be placed on the beaches. But the same is not true for the rest of the country.

Beach sand from ship channels is still being thrown away along all of our coasts. Over 20 million cubic yards of sand have been dumped in deepwater since 1975 at Mobile Pass in the Gulf of Mexico. Over a million cubic yards of dredged sediments, mostly sand, from the Columbia River entrance are dumped offshore in the Pacific every year. A quarter of a million cubic yards of sands dredged from Fairport Harbor have been dumped in Lake Erie in the last several decades.

The solution to this problem is "sand bypassing." Sand bypassing is moving the beach sand across the inlet to replace the natural movement of sand along the coast that the ship channel has to interrupt. Sand bypassing is sustainable development. Usually sand bypassing is just having the dredge pump the sand to the downdrift beaches. The jetties at some inlets were specifically designed to reduce the cost of sand bypassing. At a very few inlets in America, fixed bypassing plants are built to make this easier and cheaper. These are small, unique dredges that suck sand from the beach on one side of the inlet and pump it to the other side of the inlet. The Indian River, Delaware bypassing plant moves about 100,000 cubic yards of sand per year across that inlet. It has successfully stopped the beach erosion on the downdrift side.

Sand is trapped on the south side (left side of this photo) of the Manasquan Inlet jetty in New Jersey and the beaches on the other side are eroded. The dominant direction of longshore sand movement is to the north (to the right) along this portion of the New Jersey coast.

Bypassing has been tried occasionally at many locations. The problem is that we haven't done a very good job of it. As Mike Walthers, a coastal engineer from Vero Beach, Florida, puts it; good bypassing is more than "just taking the sand out of the channel and putting it on the beach." Often the dredged sand is just placed in the most convenient location instead of in the best location to preserve the adjacent beaches. The "least cost" disposal of sand is often the legal requirement. The problem is that the "least cost" is based on maintaining the channel and not on the "least cost" for maintaining the downdrift beaches. There are policy problems because the organization responsible for maintaining navigation is usually not really responsible for maintaining the beaches. Technical problems also complicate sand bypassing efforts. It often costs more money to place the sand on the beach because most dredging equipment is designed for deepwater disposal. Also, some of the sand in the deep ship channels is mixed with fine silts that we don't want on the beach. And since sand moves both ways along many of our coasts, it is not always clear where the sand should go. Because of the two-way nature of the wave-driven "river of sand" moving along our coasts, the best place to put the sand so that it does not get trapped against the jetties can be several miles down the beach in either direction.

In the big picture, sand bypassing is needed to save America's beaches. We must restore the movement of sand along and to our coasts. The details of the sand bypassing operation at each channel must be worked out. The sand needs to be placed where it avoids downdrift erosion by re-establishing the "river of sand." This will usually cost more money than a simpler approach but it is needed. Ship channels are important to society but so are sandy beaches. We can have both safe navigation channels and healthy beaches: indeed, for the economic strength of the nation, we must.

Beach sand robbery by groins

Groins are one of the most common and yet most hated approaches to reducing beach erosion in America. Groins are rock or wood walls built perpendicular to the shoreline to trap sand and stabilize the sandy beach. Although many Americans call these "jetties," that term is usually reserved for their cousins built to stabilize inlets. Groins on beaches are hated primarily because their sand robbery is so obvious! Even the casual observer can recognize that sand builds up on one side of the groin and erodes on the other. It is said that groins "rob from Peter to pay Paul." If all sand thieves were this obvious, our beaches might be much healthier today!

One highly publicized case of beach erosion caused by groins is Westhampton Beach, New York. In the 1960's a groin field was constructed in the eastern half of the town on the Atlantic Ocean beaches of Long Island. The dominant direction of longshore sand transport, the river of sand, is to the west along these beaches. A series of 15 long, tall, rock groins were built. No beach nourishment was included.

And the groins worked too well! They trapped so much sand that the downdrift beaches were starved and eroded and never recovered. During every major storm, the starved beaches experienced more overwash than the eastern beaches and the houses suffered more damages than their neighbors.

The media frequently showed pictures of the storm damages here while discussing the dangers of living along a dynamic barrier island. Was this damage due to the natural hazards of living along the coast, or due the man-made hazard of living downdrift of a groin field that stole a lot of beach sand? Obviously, the groin field caused the erosion. Yet, it was portrayed in the media as primarily a natural erosion problem

Groins caused severe beach erosion at Westhampton Beach, New York. The rock groins trapped sand on the beaches in the background and starved the beaches in the foreground.

and Westhampton Beach was repeatedly used as an example of the folly of trying to build houses on barrier islands. The western part of the town, now its own town called Westhampton Dunes, was restored by beach nourishment in the 1990's after a long court battle that ultimately led to a legal settlement that acknowledged the effects of the groins. While the legal and policy implications of the Westhampton groin field case are often still debated, the technical issues are clear and were clear from the start: the groins trapped sand and starved the downdrift beaches.

Groins, structures perpendicular to the shoreline, have been built along many of America's beaches

The same sort of trapping has occurred to some extent at almost every groin ever constructed. Indeed, groins are built to trap sand and "stabilize" the beach. Groins have a long history of use in America. They were the primary beach erosion response until the 1960's. Today it is difficult to legally build new groins because of their obvious sand robbing record. However, many American coastal communities have them and need to make decisions about their future. Rich Weggel, a professor at Drexel University, points out that "groins are both a boon and a bane" because they do indeed stabilize part of the shoreline when designed correctly.

Some type of groin or breakwater structure is occasionally used with beach nourishment to retain the new sand for a longer time. The idea is to "extend the life" of the beach nourishment. Well-designed structures can be particularly effective at the ends of beaches beside inlets. The nourished sand is wanted on the beach and not in the inlet. These structures; some are groins, some are jetties, some are offshore breakwaters, and some are combinations of groins and breakwaters called "t-head" breakwaters or groins; have to be very carefully designed to ensure they don't cause some beach erosion elsewhere.

Beach sand robbery by inlet jetties

Offshore dumping of dredged sands is only one of the ways ship channels hurt beaches. Inlet jetties are even bigger thieves of beach sands. Jetties are rock walls built to keep sand out of ship channels. They do this by blocking the movement of sand along the beach and by concentrating the outgoing ebb-tidal currents to "jet" the sand out to deeper water. That is how they got their name. And they have done such a good job that they have robbed America's beaches of hundreds of millions of cubic yards of sand.

One of the most dramatic examples of beach erosion caused by jetties is Assateaque Island, Maryland. A triple-whammy of effects combined to cause severe beach erosion. There was some trapping of sand on the updrift side of the jetty like there is at a groin: there was some dredged sand thrown away in deepwater: but there was also trapping of sand in a new underwater shoal. The three types of effects combined to trap enough sand to destroy the downdrift barrier island.

A hurricane hit Ocean City, Maryland in 1933 and caused a new inlet to form through the southern end of the town. Local fishermen immediately liked the new inlet since it saved them time getting from the bay to the sea. The new inlet was stabilized with jetties, and as they say... the rest is history. The dominant direction of longshore sand movement, the river of sand, along this coast is to the south. The beaches on the north side widened in the first few years. The jetties also did their job and "jetted" sand offshore. They jetted so much sand offshore that a new, large, ebb-tidal shoal developed seaward of the jetties. This shoal began to develop in 1935.

Location of new sand shoal

By the mid-1960's it included about 6 million cubic yards of sand. By the 1990's, it included about 13 million cubic yards of sand. The beaches on the downdrift, south side, Assateague Island, began to erode in 1933. For the next seven decades, continuing to this day, the Assateague side overwashed more frequently and more severely during storms than the Ocean City side. Each overwashing storm moves sand from the Atlantic side of Assateague to the bay side of the island. Essentially, the island is "rolling over" itself. The Atlantic shoreline has moved over 2000 feet landward and the island has migrated more than its width.

Inlet jetties are America's biggest sand thieves. The Ocean City, Maryland inlet and jetties shown here have caused severe beach erosion for miles on Assateague Island (the bottom island in this picture). A tremendous amount of sand is now stored in a new ebb-tidal shoal underwater near the tip of the jetties (arrows). This sand, which came off the beaches, was "jetted" out here by the jetties and the tidal currents.

Much of the erosion on Assateague was initially blamed on storms. But blaming storms for beach erosion is like blaming gravity for plane crashes! There was something else wrong. At Assateague, the sand in the new ebb-tidal shoal at Ocean City Inlet was a major withdrawal from the sand budget of the area. And the beaches of Assateague paid the price. Sand that was moving south along the Ocean City beaches on its way to the Assateague beaches didn't get there because it got trapped in the new shoal. Thus, the jetties, which improved the safety of the boating, caused erosion along the adjacent beaches.

Nearly every inlet jetty in America has trapped beach sands. There are other ways that inlet jetties trap sand. Sand is trapped on flood-tidal deltas in the inner bay at some inlets like Sebastian Inlet, Florida. Sand is trapped in widened beaches on both sides of many jettied inlets since waves drive sand both ways along most coasts. The jetties provide some sheltering from waves and the beaches widen on both sides near the jetty and often erode a little farther away because of this trapping. Sometimes, the erosion is greatest several miles from the jetties. Sand is trapped on both sides of the Yaquina Bay jetties at Newport, Oregon. The Columbia River jetties have trapped so much sand, you can't even see the ocean from some of the old gun emplacements! Conneaut Harbor, Ohio has trapped nearly 850,000 cubic yards of sand and contributed to downdrift erosion that extends to Pennsylvania a few miles downdrift.

The trapping of sand at each inlet jetty has some unique aspects but all follow the basic sand budget idea. The equation is simple: a cubic yard of sand trapped by the structure or thrown away by dredging is a cubic yard of beach erosion somewhere else.

Southern California's beach erosion is partially caused by river engineering. Many rivers that flow directly to the sea have been lined with concrete to pass floodwaters quicker. This has stopped sand from eroding from these rivers' banks and moving to the beach in large floods. Also, sand is mined from some of these river beds near the coast.

Beach sand robbery by western river structures

On the west coast of the United States, dams and other engineering on the small coastal rivers rob sand from the beaches. Because of the geology of this coastline, sand from these rivers naturally feeds the beaches during every major rainstorm. Dams reduce the amount of sand reaching the coast two ways. One, they trap it. Two, they reduce the peak floods that move the sand to the beach. It is difficult to estimate the reductions caused by the dams but, in southern California, over 80% of the watersheds that drain to the beaches are now behind dams. Lining coastal rivers with concrete walls also reduces sand moving to the beaches since it prevents the erosion of sand bars and sandy riverbanks. These are a primary source of sand for the beaches of southern California.

The Matilija Dam has trapped over 6 million cubic yards of sand, gravel, and silt that was on its way to the beaches of Ventura, California. The dam was built in 1948 sixteen miles up the Ventura River from the beach. It was originally built for water

supply and flood control reasons - the same reasons many dams have been constructed throughout the country. The reservoir behind the Matilija Dam has filled in with so much sediment that its water storage has been reduced by 90%. The reservoir is half its original size in area and the depths are much shallower. The dam has also probably hurt the steelhead fishery because of gravel trapping and river flow changes. This dam is being removed because of efforts by a broad coalition of interests. One of the important details of this removal plan is the question of what to do with the

Dams on many of the coastal rivers of the Pacific coast have contributed to beach erosion. The dams trap sand that was on its way to the beach and also reduce floodwater peaks that would have eroded more of the downstream banks onto the beaches. This is the Matilija Dam upstream from the Ventura beaches.

trapped sediments. Should we move them to the beach artificially or let them move there via floods? In any case, it appears that the pathway of sand movement to the beaches is going to be partially restored on the Ventura River. But the problem will remain in most other western, coastal watersheds.

Dams are probably not a significant beach sand thief in much of the country. The geology of the barrier islands of the Atlantic is very different than the geology of the beaches of southern California. The sand in the barrier islands is sand the eroded out of the mountains many hundreds of thousands of years ago and the rivers are no longer supplying much sand to the coast.

Beach sand robbery by sand mining

One obvious sand thief is mining of sand from the beach or from the areas that feed the beach. California has a long history of doing both! Sand is a valuable construction material. For over a half a century, sand was mined directly off the beach in Monterey Bay. One mining operation used a dragline to scrape the beach through the surf zone. Longshore sand transport would move sand back into the area every night. Orville Magoon tells the story of how the mine operator called the federal engineers and asked if they could fix his beach erosion problem! When they suggested that he was causing his own problem, his response was "no, there's plenty of sand out there. It comes back every night!" About 1990, that mine lost its permit but the damage to that stretch of Pacific shoreline was permanent. California has also had many sand mines in the river beds within a few miles of the beach. These are probably causing erosion of the ocean beaches that are downstream. Several sand mines are still operated in the sand dunes right behind the beaches.

Beach sand mining has caused millions of cubic yards of beach erosion. Sand was mined directly from the active beach face in Monterey Bay, California until about 1990. The active mine shown here is in the sand dunes behind the beach.

There are many places in America where some indirect mining of beach sand has occurred and is still occurring today. At many inlets throughout the country, dredged sand is placed in upland disposal sites and then hauled away by truck to "free up more disposal capacity" for future dredging needs. Near many inlets, dredged sand has been placed in upland areas to raise the land elevation. The same simple equation holds here: a cubic yard of sand mining, whether direct or not, is a cubic yard of beach erosion somewhere.

Sand has been indirectly mined from many American beaches by decisions at nearby inlet boat channels. This sand is being hauled away by truck years after it was first dredged from the channel in Perdido Pass, Alabama and placed in this upland "disposal area."

Beach sand robbery by seawalls

Seawalls can rob beaches of sand just by doing their job. Seawalls don't help beaches. They just protect upland property. Seawalls are usually built on shorelines that are already eroding and they don't address the sand budget deficit in the area that is causing the erosion. The seawall doesn't really cause the erosion: it just doesn't fix it. Thus, unless something else changes, the beach in front of the wall continues to erode and many seawalls eventually have no dry beach in front of them. There are some places where the sand budget is such that beaches exist seaward of a seawall. Many of these are where beach nourishment has restored the beach. Other places are where the longshore sand transport is balanced or where there is sand that has migrated in from offshore. This can occur near inlets where shoals from the inlets weld onshore occasionally.

Seawalls don't help beaches. They only protect upland property. This massive seawall was built when the elevation of Galveston was raised after the Hurricane of 1900. It has done its job of protecting the city from storm surges. But the beaches have eroded for other reasons.

Seawalls can rob beaches of sand by not allowing the bluff or dune they protect to erode and feed the adjacent beaches. The amount of sand thus robbed from the beach is related to the erosion rate and the material in the bluff. Many eroding coastal bluffs have only a small percentage of sand size material. Some are 100% sand.

Coastal bluff erosion is reduced when there is a wide sandy beach at the base of the bluff. The absence of a sandy beach at the base of a bluff means the waves are free to pound on the bluff more frequently and with more force. For example, the rate of bluff erosion increased downdrift of a long jetty on the Lake Erie shore at Fairport Harbor, Ohio. The bluff erosion rate decreased on the updrift side of the jetty because it has trapped a wider beach while starving the downdrift side. The jetty blocked the sand movement along the beaches and thus influenced the bluff erosion rate.

Seawalls are outlawed in some states and encouraged in others. The number of seawalls in the surf has dropped in New Jersey and Florida for the past decade as they have been buried behind beach nourishment sands. But 80% of Ohio's Lake Erie shoreline is armored with some form of seawall and in the metropolitan areas the percentage of walled shorelines is 96%. More and more of the California coast is being armored with seawalls. 20% to 25% of the coast between San Francisco and San Diego is now armored! Since seawalls don't save beaches, the obvious fate of California's coast is to turn into a long series of seawalls with no sandy beaches unless some other approach like beach nourishment is embraced.

Seawalls that stop bluff erosion at the coast also indirectly rob the beach of sand. Some of the beach's sand eroded from the bluff. Much of the California coast is becoming armored with seawalls. This rock seawall at San Clemente, California protects the railroad tracks but now there is no beach here except at lower tides.

Not many seawalls would be needed if we were willing to back off the beaches and bluffs farther. Thus, "backing off" is one solution to beach erosion problems. It is probably the only way that an individual beachfront property owner can widen his or her own beach. It should be considered whenever and wherever possible. "Backing off" doesn't address the erosion, but only the problem caused by the erosion. The beaches of America would survive just fine if they were allowed to move freely without hitting seawalls and if the sand robbery were stopped.

The history of many coastal states includes policy battles to set new construction back away from the water's edge. The hazards of encroachment, building too close to the sea, have been obvious for years. The hazards are to life and property but also to the beach. The concept behind setbacks is to provide a "buffer zone" between the surf and the buildings. This concept is a good one whether the buffer is for long-term erosion or storm damage reduction. There is a common value to setbacks because if everyone sets back an equal amount, then everyone gets the benefits of oceanfront views and the benefits of a healthy buffer zone.

Another related policy option is "retreat" – moving buildings back as the beaches erode. While this could work, from a technical perspective, to save America's beaches, it has rarely been implemented because of legal reasons. Cries of "planned retreat is insanity" are coming from California today. Cries of "no retreat, baby, no surrender" came from New Jersey ten years ago. And cries of "retreat, hell" came from Florida ten years before that! Retreat just isn't happening in America for many legal, societal and political reasons.

This seawall is disguised to look like the natural bluff at Solano Beach, California. The man in this photo is looking at a reinforced, tied-back, concrete seawall with fake cracks and color to match the native bluffs here. The native, eroding bluff is on the left side of the photo and at the very far right. Note the narrow beach on a calm day.

In California and many other places in America, we are building walls along shorelines that are being starved by sand thieves. We aren't stopping the sand robbery and we aren't backing off effectively. The result is that we gradually, but certainly, are destroying our beaches. There is only one other approach that both protects upland property and preserves a sandy beach – beach nourishment.

Chapter 5

"Designer Beaches" —
Beach Nourishment Engineering

The wide, sandy beach stretched out before Megan and Kate. It was covered with people for as far as they could see. Their weeklong family vacation had just begun and they had just raced to the beach as quickly as possible. The beach the youngsters looked at was spectacular. For the next week, this would be their kingdom to "lay out" in their new bikinis, ride waves on their boogie boards, build sandcastles with their cousins, and go on long walks to check out the boys. In short, it was paradise. And it was a nourished beach. The girls didn't know they were looking at a nourished beach and they didn't care. But without the beach nourishment, the girls would have been looking at a much different scene. They would have had little or no space to lay their towels or build their sandcastles. Even if they found a place, it would have been in the shadow of a seawall. Without the beach nourishment, they could not have walked along the beach except at low tide because of the water hitting the walls. The beach, as they knew it and loved it, had been saved by beach nourishment.

Dozens of America's best beaches have been saved by beach nourishment. Beach nourishment is the placement of large amounts of good quality sand on the beach to widen it. Nourishment is one of the engineering remedies to beach erosion that has been widely developed in the last few decades. Knowledge from geology, oceanography, ecology, and civil engineering are combined with high-tech computer simulations and old-fashioned judgment based on experience. And every beach nourishment design is unique.

In fact, now, a good beach nourishment project is rather like a designer dress. You pay to get the look you want designed for your specific beach's attributes and your specific desires. The size of the beach, the shape of the beach, the look and feel of the beach can all be manipulated to some extent. And, like some designer dresses, some beach nourishment look really, really good.

Beach nourishment can work well too. The National Research Council of the National Academies of Science and Engineering concluded in 1995 "beach nourishment is a viable engineering alternative for shore protection and is the principal technique for beach restoration." It can widen beaches and "provide protection from storm and flooding damage." Over 300 miles of America's most popular beaches have been saved by nourishment. In the big picture, beach nourishment is one way to bring back many of the beaches of America and save them for future generations.

Miami Beach has shown that beach nourishment can work to restore beaches. Much of the beach was gone before the 1977 beach nourishment (upper photo) and some beaches are over 600 feet wide today (lower photo).

"If sand is the problem, then sand is the solution"

The beach of Miami Beach was saved by nourishment. In 1977, there was no beach! And in a city that prided itself on and based its tourist economy on its beautiful beach, this was a serious problem. Waves hit the seawalls and groins in front of oceanfront hotels every day. Tourists and citizens couldn't even walk along the beach because there was not enough sand. The problem was partly due to unwise encroachment and partly due to sand thievery at several ship channels. For over ten years, the solution to the problem – no beach – had been debated fiercely. It was obvious that bigger seawalls would not restore the beach and it was obvious that moving the hotels back was not feasible. Critics said beach nourishment wouldn't work and would "wash away in the first storm." It would be very expensive and would be "throwing dollars into the sea."

The history of beach nourishment in the United States goes back further than Miami Beach. City leaders placed sand on some beaches many decades ago to make them wider. In fact, some of the most beloved old beaches in America are man-made, nourished beaches. The beaches of Coney Island, New York were built in 1923. The "Baywatch" beaches of Santa Monica Bay in Los Angeles County, including Santa Monica, Venice Beach, Dockweiler Beach and Manhattan Beach were widened by man in the 1940's and 1960's. The Great Lakes and Gulf of Mexico coasts also have some very old, very popular, man-made beaches. The pioneering lakefront park that Chicago built in the 1920's and 1930's included over a dozen sandy beaches for recreation. These beaches are covered with thousands of people every July 4th. Biloxi, Mississippi created their signature, 26-mile long, sandy beach by pumping sand from offshore in the 1950's, renourishing it once in the 70's, once in the 80's and again in 2001.

These older man-made beaches can be thought of as "primitive" or "pioneer" beachfills since they were built before the development of many of the modern engineering tools for beach nourishment. We now have much more experience with restoring American beaches. Many of the technical questions about beach nourishment that were open questions in the 1970's now have clear answers.

At Miami Beach, the results are very clear. Beach nourishment has worked. Today, the beaches of Miami Beach are very wide. In some places in the South Beach area of Miami Beach, the beaches are almost too wide! It is over 600 feet to the water. On average, the Miami Beach beaches are 300 feet wider today than in 1977 thanks to the 9 million cubic yards of sand placed there as beach nourishment.

The beach nourishment debate has raged through hundreds of American beach communities much as it did at Miami Beach in the 1970's. And more and more American coastal communities are making beach nourishment work for them. The reason, as Tim Kana, a coastal geologist from Columbia, South Carolina says, is that beach nourishment is the best of the available choices for saving the sandy beach. There are really only three general ways to respond to beach erosion: retreat by removing oceanfront buildings and roads, armor the coast with walls to protect the oceanfront buildings, and beach nourishment. The first way, retreat, has proven to be terribly difficult, legally and politically, in America. The costs of retreat are just too great. And the second way, building seawalls, essentially dooms the beach.   Thus, beach nourishment has become the solution of choice for most American beach communities. As Bob Dean of the University of Florida says, "if sand is the problem, then sand (beach nourishment) is often the best solution" for most communities that are faced with a beach erosion problem.

Beach nourishment protects property. The Ocean City, Maryland beach nourishment project saved millions of dollars by reducing the damages to public and private property in the "Perfect Storm" of 1991 and other major storms the next year.

An added benefit is that beach nourishment protects buildings during storms. Nourishment sand provides a "buffer zone" between the sea and the buildings. When Hurricanes Dennis and Floyd hammered the North Carolina coast back-to-back in 1999, almost 1000 homes officially were either "damaged" or "threatened." Amazingly, none of those 1000 homes was behind the two largest nourished beaches in the area, Wrightsville Beach and Carolina Beach! Those two towns have had their beaches and sand dunes maintained with nourishment since 1965. Homes right at the ends of the nourished beaches, however, were damaged. The same thing is found other places after every storm. The Miami Beach nourishment has survived a number of hurricanes and tropical storms. The sand "takes the beating instead of the seawalls and buildings" says Brian Flynn of Dade County. Federal government economists estimated that the 1988-1991 nourishment at Ocean City, Maryland reduced storm damages by between $52 and $160 million in the "Perfect Storm" of 1991 and several other severe northeasters that followed within the next year. The mayor of Ocean City said that the new beach "saved" the town.

Beach nourishment built this Ocean City, New Jersey beach seaward of the boardwalk.   The upper picture is from 1986 and the lower is from 2001.

"Beach nourishment is a journey, not a destination"

Most beaches that are nourished will need to be "re-nourished." Some of the nourished sand will eventually move down the beach to adjacent beaches or inlets. Nourishing a beach is like painting a house – you are probably going to want to do it again. Painting is a part of house maintenance that is done periodically to protect the investment beneath the paint (the house) and to make it nice for our enjoyment. Likewise, beach renourishment is part of beach maintenance that is done periodically to protect the investment behind the beach (buildings and roads) and to make it nice for our enjoyment. If we do a good job of painting the house; with scraping, caulking and using good quality paint; the paint job will perform better and last longer. Likewise, if we do a good job of beach nourishment, with a sound coastal engineering design based on a good monitoring program and including enough good sand, the beach nourishment will perform better and last longer.

Nourished beaches can apparently last forever if they are well maintained! Miami Beach has proven that in spite of the original nay saying. Delray Beach, Florida has one of the longest-running, regular beach nourishment programs in the country. This small, south Florida town on the Atlantic Ocean built their original beach nourishment in 1973 with 1.6 million cubic yards of sand along 2½ miles of coastline. Much of that coastline had a seawall and no dry, sandy beach at that time. Beach nourishment restored the beach and it has been there ever since. Five years after the original beach nourishment of Delray Beach, about 40% of the sand had moved to the beaches of the two neighboring towns. Renourishments in 1978, 1984 and 1992 added another 2.9 million cubic yards of sand. Delray Beach was scheduled for renourishment in 2002. The total amount of sand on the beaches and dunes of Delray Beach has increased due to the nourishments. The beaches are over 150 feet wider on average than they were in 1972 and all of them are much higher. A sand dune ecosystem has developed right on top of an old, failing seawall! The beach and dune restoration has been so successful at re-establishing a natural-looking coastline that many of the newer residents of the town do not even realize that the beach and dune are the result of a nourishment program! This has been a typical comment at many well-built beaches. It's really the ultimate compliment to the beach engineers and designers.

"Beach nourishment is a journey, not a destination," says Tom Campbell, one of the coastal engineers for the Delray Beach beach nourishment. Future renourishments are part of the town's general planning and budgeting process but the details of the design change with time. A beach in an ongoing nourishment program begins to retain more and more of its previous nourishment sands. Also, the environmental and cost constraints of the design will change through time. One critical aspect of any beach nourishment is post-construction monitoring. Careful, expensive measurements of where the sand goes and why it does are vital to the design of the overall beach restoration program. At Delray Beach, design changes have been based on the monitoring results. Essentially, the beach nourishment "journey" has saved the beaches of Delray Beach.

Beach nourishment can apparently last forever if it is maintained. This dune and beach at Delray Beach, Florida have been here for 30 years (upper two photographs from 2001) because of beach nourishment, dune plantings and subsequent renourishments that were based on careful monitoring and design. Prior to nourishment, there was a failing seawall with no beach or dune (lower photograph).

The time between beach renourishments, the renourishment interval, varies from beach to beach. The interval is related to the age of the nourishment project, the local wave climate, the overall length of the nourishment area, the amount of sand used, the nourishment sand grain size, the background erosion rate and other factors. The time between renourishments often increases with age. For example, at Delray Beach, the time between renourishments increased from 5 to 6 to 8 to 10 years. Shorter intervals are common when not enough sand is used or when the sand grain size is smaller than the native beach sand. Shorter intervals are common on beaches that have more wave energy. Because of these differences, it is difficult to call any renourishment interval "typical." Experience with renourishment intervals ranges from 2 to 25 years with many being 6 to 10 years.

Less sand is usually needed as time goes on

Renourishments often use less sand than the original nourishment because much of the original nourishment sand is still there. This is important because the long-term sustainability of beach nourishment is being questioned if more and more sand will be needed in the future as sea levels continue to rise. Certainly, if the rate of sea level rise accelerates dramatically because of global climate change, the beaches are going to respond very differently than they have for the past 4000 years. But our experience with beach nourishment in the last 40 years shows we will need less sand to maintain our nourished beaches. At Miami Beach, all the renourishments in 25 years have totaled less than 20% of the initial sand needs. Five miles of Miami Beach have yet to need any renourishment at all!

Bob Dean of the University of Florida contends that the past 30 years of beach nourishments in Florida were a "catch-up phase" responding to the effects of a century of sand thievery and encroachment. Now we are entering a "maintenance phase" that will need much less sand. Indeed, the total amount of beach nourishment in the last three decades of the 20th century along the southeast Florida coast is about equal to the total amount of sand dredged from the ship channels and dumped offshore in the first seven decades.

Many nourished beaches in America are directly downdrift of inlets that have robbed the beach sand. These beach nourishments have restored the beaches destroyed by the sand thieves at the inlet. They've brought back the beaches.

"Bigger is better" for beach nourishment performance

Bigger beach nourishments last longer. This means both using more sand and extending it farther down the beach. Sand that moves from one spot on a long, nourished beach just moves down the beach to another spot on the same project. This concept is being used more often. The 1998 Panama City, Florida beach nourishment was almost 17 miles long. About 8 million cubic yards of sand was placed along the resort town in the Florida Panhandle on the Gulf of Mexico. The nourishment restored a beach that had been severely narrowed by a combination of encroachment of the buildings and sand robbery at the updrift inlet. Along the worst sections, there was no dry beach, just a seawall in the surf before nourishment. The oceanfront buildings were damaged during small storms because of the narrow beach. After nourishment, the beach survived Hurricane Georges and the damages to oceanfront buildings were much less than would have occurred without the nourished beach.

If beach nourishment can work at Sea Bright and Monmouth Beach, New Jersey, it can probably work anywhere! The two northern New Jersey towns are part of one of the largest beach nourishment projects in America. A new sandy beach was created

Beach nourishment restored this Gulf of Mexico beach at Panama City, Florida. A white sand beach now buries the seawall (shown in 1997 to left and 1999 above).

seaward of miles of seawall that had been in the surf for over 70 years. There was a seawall with no sandy beach. The water was up to 20 feet deep immediately seaward of the wall and waves crashed against the wall regularly. Because of the pre-existing sand deficit, this nourishment used about 10 times as much sand per foot of beach as the Panama City project. Today, there is a dune and wide beach seaward of the old Monmouth and Sea Bright seawalls. Beach nourishment saved, or more accurately, brought back, these beaches that had been missing for generations. Originally built from 1994 to 1996, portions of the beach nourishment were scheduled for renourishment in 2002. Thus, the renourishment interval for this beach has been 7 years. The design prediction for the first renourishment interval was 6 years. There were several mild years in the wave climate after construction and so renourishment was postponed.

The "bigger is better" concept is also one reason why beach nourishment is typically a government responsibility. A single beachfront owner doesn't have a long enough property to maintain a nourished beach. Waves will remove all of the sand from a very short nourished beach very quickly. This is leading to more and more regional approaches to nourishment. Several adjacent beach towns can work together to maintain their beaches more effectively than they can independently.

Predicting the fate of beach nourishment

Many important questions relate to the fate or performance of beach nourishments. How long it will last? Will it wash away in the first storm? Where will the sand go? In spite of tremendous complexities in the surf zone, experienced beach engineers can predict the fate of beach nourishments fairly well.

When sand moves away from where it is placed is often called a "loss" of the sand. But, the sand isn't really lost at all. Monitoring shows us exactly where it is. At Delray Beach, Florida, all of the "lost" sand was "found" on the beaches of the neighboring towns. Many beach nourishments become "feeder beaches" for the nearby coast.

Nourished beaches feel the same forces as natural beaches. The wave-driven littoral drift, or "river of sand," will always move nourishment sands down the coast. The movement is often in both directions, left and right, but with some dominance in one direction depending on the wave climate. Around Delray Beach, which is less than 3 miles long, the beaches have gotten wider for about 7 miles! In other words, the beaches have widened for about 2 miles on either side of the beach nourishment. About 60% of the nourishment sand that left Delray Beach went south to widen the beaches of Highland Beach and 40% of it went north to widen the beaches of Gulfstream. The beaches in these two towns have widened over 70 feet thanks to the nourishment in Delray Beach. All nourished beaches do the same thing.

Hilton Head, South Carolina has saved its beaches with beach nourishment.

Because wave-driven sand movement, the "river of sand," extends many miles down the coast, political town boundaries are artificial in terms of the physical forces shaping the beaches. Much of the sand from the Atlantic City beach nourishments moves from those beaches to the beaches of Ventnor and Margate, towns on the same barrier island immediately south of Atlantic City. Thus, this sand was not really lost to the overall beach system since the sand widened the beaches downdrift.

Several methods have been developed for estimating the movement of sand down the beach. Since nourishment sand often spreads out in both directions away from where it is placed, engineers use the analogy of the "diffusion" process. Diffusion concepts explain many things in physics. A rough analogy is the change in the location of a group of children getting off a bus. They all get off the bus at the same point but a few minutes later, they are spread all over the place and the number of kids right near the bus decays exponentially with time. Beach engineers estimate the spreading out of sand in the design of beach nourishment projects. The beach sand "diffusion" is described by equations and the result is called a "model." And the "design life" or renourishment interval can be estimated with that model. The diffusion model has done a good job of predicting the fate of the sand at Delray Beach, Florida and many other beach nourishments.

Hilton Head Island, South Carolina has inlets, not adjacent beaches, at both ends of its beach. Sand that moves out of the beach nourishment area often moves into the inlets. Thus, from the perspective of the beach, these are much more like true "end-losses." The inlets along this part of the South Carolina coast store tremendous amounts of sand in the ebb-tidal shoals. These shoals affect the wave climate on the island beaches as well as provide a source for the beach nourishment sands. All beach nourishment projects need to be designed with a sound understanding of the local coastal processes that move sand around.

Kevin Bodge, one of the design engineers for the 1990 nourishment of Hilton Head Island, predicted the amount of sand that would stay on the resort's beaches. He estimated that 75% of the sand would be there after four years; and 50% after seven years. The "lost" sand was going to move off the beaches into the inlets. The predictions were almost exactly right! The actual amount of sand was measured with a monitoring program that surveyed the beach and dunes. After four years, 75% of the nourishment sand volume was there. When the beaches were renourished in 1997, 58% of the 1990 sand was still on the beaches. The engineering predictions are not usually this good and significant uncertainty abounds. But with good judgement it appears that these predictions can work well even in complex situations like Hilton Head.

Bob Dean of the University of Florida has compared the predicted and actual performance of eight different beach nourishments in Florida. The actual amount of sand remaining on the beach through time was within 30% of the predictions on average. Thus, these are not perfect predictions but they are pretty good ballpark estimates. "I think it will provide some confidence to officials embarking on renourishment projects for their communities, although we would like to do better," Dean says. These sorts of predictions should be thought of as the best available, rough estimates. They can be thought of as "diagnostic tools" for the decision-makers that must be used with sound judgment based on experience.

"Hotspots" drive renourishment

Beach renourishment decisions are often driven by so-called "hotspots." The "hotspots" are the areas where the nourished beaches are narrowest. When the nourished beach erodes back to a point where <u>any</u> part of the beach is narrow or non-existent, the public perception is that it is time to renourish. All parts of a single beach nourishment do not behave the same. Some areas will lose beach width faster than others.

Even the worst "hotspots" on nourished beaches often are better beaches than before beach nourishment. This part of Myrtle Beach, near its north end, has some of the narrowest beaches in the nourished area today (lower photo from 2001). However, prior to nourishment there was no dry beach at high tide (upper photo from 1987). Also, note that the condominium here was under construction in 1987 in spite of the condition of the beach! "Backing off" wasn't accomplished as a solution to saving the beach but beach nourishment succeeded.

A good beach designer can usually predict the future "hotspots" in advance. The "hotspots" are usually areas that have some of the worst problems before nourishment. For example, "hotspots" are sometimes downdrift of inlets that are still trapping sand. Other "hotspots" are found at the ends of islands near inlets where littoral drift moves sand into the adjacent inlet and the beaches are influenced by inlet shoal fluctuations. Two of the most severe "hotspots" for the Myrtle Beach, South Carolina beach nourishment project are at the ends near the adjacent inlets. Another common "hotspot" location is where the beach is protruding seaward compared to adjacent beaches. The 32nd Street area in Miami Beach has been one of the "hotspots" that hasn't stayed as wide as the other beaches. The shoreline and the line of buildings bulge slightly farther seaward than the adjacent beaches. This allows waves to move the nourishment sand more quickly away. Other typical "hotspot" locations are

Sands for nourishment usually come from offshore sand deposits. Many of the Pinellas County, Florida beaches, in the St. Petersburg/Tampa area have been saved with beach nourishment from offshore sand sources.

where buildings encroached onto the beach before nourishment. This problem is worse where waves have been regularly hitting a seawall before nourishment. There is a greater sand deficit that must be overcome. The beach is overly starved before nourishment and more sand is needed.

Some of the aspects of beach nourishment and renourishment are related to the political decision-making process that pays for it. Mike Walthers, a coastal engineer in Vero Beach, Florida, says the design of beach restoration projects is like a three-legged stool. The legs are the technical know-how, the financing, and the public perception. Without any of the three legs, the stool collapses. Public outreach through community meetings, press releases, signs and web pages are a critical part of good beach nourishment design.

Where does the sand come from?

The source of sand is one of the critical parts of beach nourishment design. The "quality" of the sand controls the aesthetics, the cost, and the physical and ecological performance. The perfect sand for beach nourishment would be sand that is the exact same as the native beach sand. The match should include grain size and distribution, shell fragment content, percent of fines (usually zero), color and mineral content.

However, we can rarely match Mother Nature perfectly. It appears that it is better to use sand grains that are too big than to use sand grains that are too small. Nourishment with finer-grained sands has resulted in very disappointing performance. The finer-grained sands wash away too rapidly and shift too far offshore. Some of the negative public perception of beach nourishment is probably due to source sands that were not of a high enough quality.

Many of the best sand sources for nourishment are offshore, underwater sand deposits. They often are a lot like the native beach sands and it can be easy to mine them. These deposits can have different geologic origins. Some beach nourishment sand sources are ebb-tidal or flood-tidal shoals, some are ancient offshore sand deposits and some are navigation channels that need dredging anyway.

This beach on Key West, Smathers Beach, was nourished with sand hauled by truck from over 200 miles away!

The 1994-1996 northern New Jersey beach nourishment used sand from two areas about 4 miles offshore. The source areas were underwater sand ridges that are probably the remnants of a beach and dune system that was there over 10,000 years ago and was drowned offshore during the most recent sea level rise. Offshore sources have been used for beach nourishment along all four of America's coasts.

Ebb-tidal shoals have been mined for beach nourishment in some parts of the country. These are the shallow shoals outside of the tidal inlets. The 1997 beach nourishment of Ocean City, New Jersey mined the ebb-tidal shoal in Great Egg Harbor Inlet just to the north of the city. The 1990 and 1997 Hilton Head Island, South Carolina nourishment sand sources were parts of the adjacent ebb-tidal shoal. Ebb-tidal shoals are full of sand that came off the adjacent beaches. They are usually clean sands with grain sizes that match the beach sands. Often however, there are lenses of seashell debris in ebb-shoals.

We need to be very careful when mining ebb-tidal shoals for beach nourishment. These shoals usually shelter and feed the nearby beaches. Changing the natural shoal shape can lead to beach erosion. For example, dredging of part of the ebb-tidal shoal at Townsends Inlet, New Jersey in 1978 for beach nourishment on Sea Isle City indirectly led to beach erosion along the beaches of Avalon. Sea Isle City is on one side of the inlet and Avalon is on the other. The dredging changed the position of the main ebb-tidal channel. This changed the shape of the shoals and allowed bigger waves to reach the beaches of Avalon. Since ebb-tidal shoals often feed sand to the nearby beaches, mining them can also cause beach erosion by starvation.

Some nourishment sand comes from inland mines. The 1987 beach nourishment sand for Myrtle Beach is an example. Trucks hauled the sand to the beach. The mine was a relict sand dune system from hundreds of thousands of years ago when the sea level was higher and the beach was farther inland. The 1996 renourishment at Myrtle Beach did not use this inland mine but rather several offshore deposits. The design of beach nourishments can change.

Because the quality of nourishment sand is so important, sand searches can lead to distant and even exotic locations. The sands for the 2000 nourishment of Smathers Beach in Key West came from an inland mine over 200 miles away in south, central Florida. Fisher Island, a private island just south of Miami Beach, imported aragonite sands from the Bahamas for a small beach nourishment.

Sand "sources of opportunity" nourished some of southern California's most popular beaches. The Santa Monica Bay beaches of Los Angeles County including Manhattan Beach, of beach volleyball fame, were built from the 1940's to the 1960's. Most Americans are not aware that these wide beaches are nourished beaches!

The Santa Monica Bay beaches, better known as the "Baywatch Beaches" to the rest of America, were nourished with sand "sources of opportunity." Sand that was being dredged for some other reason was placed on the beaches. Beginning in 1947, 14 million cubic yards of sand cleared out of the coastal dunes to make room for the Hyperion Sewage Treatment Plant were put on the beach. Another 10 million cubic yards of sand dug out of Marina del Rey, the world's largest marina, were put on these beaches between 1960 and 1963. These sands spread along the coast and widened the beaches of Santa Monica, Venice Beach, Dockweiler Beach, El Segundo, Manhattan Beach, Hermosa Beach, and Redondo Beach by 150 to 250 feet. Most southern Californians don't even know that these beloved beaches are not natural! Some of southern California's other popular beaches are nourished beaches too. Huntington Beach, Newport Beach, and the Coronado Spit beaches of San Diego have all been widened by repeated beach nourishment.

The Coney Island beaches were first built in 1923 and were recently rebuilt. This is one the few beaches in America that you can ride to on a subway.

Mud, fine-grained silt or clay, should never be used for beach nourishment. Many of the problems with beach nourishment in the past have been because the grain size was too small and the percentage of "fines" in the sand was too high. Even a very small percentage of mud will make a beach fill source unacceptable for beach nourishment. The critical level depends on several things including the color of the native beach sands and the construction method. A rule of thumb is that 5% is the upper limit of mud in a beach nourishment source. Some states allow for up to 10% mud for nourishment sands. This level will be aesthetically unacceptable on most beaches. Even at levels as low as 1% to 2%, mud can look unacceptable on beaches that are typically bright white sands.

The color of beach sands can be critical. People expect to see sand on nourished beaches that looks like the sand on the native beach. This is absolutely critical in those parts of the country such as the panhandle of Florida that take pride in their "sugar-white sands." Brad Pickel, the beach management coordinator for Walton County, Florida says, "if it's not as white as our native beach, we don't want it." At other places, the color may not be as important as getting some sand on the beach. Tan, off-white, beige, gray, light-brown, brown and even sandy could all be used by different people to describe the color of the same beach sand. Like shades of color in paints, the shades of color in beach sands are infinite and their definition is typically subjective. Of course this has not stopped beach geologists and engineers from trying to quantify sand color! Typically, the Munsell color charts are used. They quantify the brightness and the hues in any color sample.

Beach nourishment has restored beaches along all of America's coasts. This Gulf of Mexico beach in Gulf Shores, Alabama was restored in 2001 - six months after the upper picture was taken and three months before the lower picture. The new beach is much wider and higher.

How beaches are built

Beach nourishment is a major construction project. Since it is done out on the beach, it usually draws crowds of curious onlookers. Typically sand is pumped from an offshore dredge through a pipeline onto the beach and then spread out with bulldozers. One of the cheapest ways to move sand is in a hydraulic pipeline. A "slurry" mixture of about 1 part sand and 10 parts water can be pumped for miles. Hydraulic dredges are like giant vacuum cleaners. They have large pumps that suck sand and water in from below the dredge. The bottom of the intake pipe has a rotating "cutterhead" that is made of hardened steel. The cutterhead acts sort of like the beater-bar on a vacuum cleaner and chews into the sand while the pump suction lifts the slurry into the barge. If the beach is within several miles of the sand source, the slurry is usually just pumped through a long pipeline. Hopper dredges, another type of dredge sometimes used, are special sand barges. They are filled up at the source, driven to near the beach, and unloaded. They are loaded and unloaded either with a clamshell or hydraulic technique. A hopper dredge can move sand long distances. Hopper dredges were used for the northern New Jersey beach nourishment. When the hopper dredge was filled with sand at the source, it sailed to a barge a few hundred yards offshore of the beach. The barge then unloaded the hopper dredge with a hydraulic pipe to the beach where it was spread along the shoreline.

Bulldozers shape the beach sands into the desired beach profile. This requires a part of the beach to be closed to the public for construction safety. Closure is often only for a day or so because the work can proceed down the beach several hundred feet per

The construction of a beach nourishment project often looks like this. An offshore ship dredges the sand off the ocean floor and pumps it through a pipeline to the beach. The pipeline discharges a mix of 10 parts water and 1 part sand onto the beach and then bulldozers are used to shape the beach when the water runs off.

day. Beach construction usually continues around the clock and the incessant "beep-beep-beep" of bulldozer's backup alarms is one of the characteristic sounds. One other characteristic of beach construction is that the seashell collecting in the new sand can be very good. If sea turtle nesting is a concern, the beach sands are tilled up to 3 feet deep to loosen them up. Finally, sand fences and dune vegetation are usually added.

"What you see is not what you get"

The "look" of a nourished beach will often change dramatically in the first year after construction. The sand will bleach in the sun and the dry part of the beach will narrow as sand moves offshore to the sand bar system. Waves re-shape a nourished beach just like any natural beach. Waves move the sand offshore and alongshore. Waves also sort out the sands. This re-shaping of the nourished beach is anticipated. And while we will never understand all of the details of complexities of the surf zone and the beach, modern coastal engineering design does account for most of this re-shaping.

Beach nourishment can be thought of as moving the beach out of equilibrium with the waves. The behavior of the nourishment can be thought of as the beach moving back toward equilibrium with the waves. The result, as Erik Olsen, a coastal engineer from Jacksonville Florida, says is "what you see (initially) is not what you get." The physical appearance of a new beach will usually change dramatically in the first 6 to 18 months.

One change is narrowing of the dry beach after construction as the new sand is pulled out onto the sand bar. The dry beach, where most of the sand is stacked by the pipes and bulldozers, is only a small portion of the whole beach. Most of the sand movement on any beach is underwater. This includes the sand bar system and can extend hundreds to thousands of feet offshore. Therefore, enough nourishment sand needs to be used so that when this expected offshore movement occurs, the entire beach is shifted seaward and the dry beach is wider than before nourishment. A nourishment designed to widen the beach 50 feet can be built such that the dry portion of the beach is 150 feet wide right after construction. Storm waves will pull this sand offshore rapidly, sometimes in the first major storm. This cross-shore sand movement is often referred to by the clunky term "profile equilibration." Instead of trying to build the beach underwater, beach engineers have found it easier to build a beach this way and let Mother Nature move the sand offshore.

Unfortunately, this expected offshore movement of sand is often improperly perceived as a "failure" of the beach nourishment project. Observers can say that "half of the new beach washed away in the first storm" or "we lost one hundred feet of our new beach already." Such statements are usually just wrongheaded. The sand didn't "wash away:" it just moved offshore as expected. And the beach width wasn't "lost:" it was never really there except in a constructed, unnaturally steep beach. Critics of beach nourishment projects sometimes intentionally use dry beach width to overstate the rate at which these beaches wash away. The sand is not "lost," it's all there underwater. The underwater portions of a beach are as important to the beach as underground foundations are to a skyscraper. The "underwater sand" is the foundation of the part we see and it has to be there. This shift of sand should not be considered losses of the beach nourishment. They are designed, expected shifts.

Many new beach sands bleach whiter in time. When initially pumped onto the beach, the sand can sometimes look terrible. But within a few weeks or months, the sand often bleaches to a lighter, native-beach-like, color. Part of the problem is that many nourishment sand sources have not been in the sun for many years. Nourishment sands bleach whiter like anything exposed to direct sunlight. This bleaching process is difficult to predict since it includes a number of physical and chemical changes to the beach sand. Some coastal geologists and engineers just put a sample from a potential borrow area in a box on their roof for several months prior to construction to evaluate bleaching! Not a particularly high-tech approach, but it works.

How expensive is beach nourishment?

Cost is a common criticism of beach nourishment. However, Howard Marlowe, a Washington lobbyist for beach communities, says, "for its relatively tiny price tag, beach replenishment is worth the investment – for our economic infrastructure, environment, and the beauty of coastal regions."

There are no "typical" costs for beach nourishment. Each beach design is too unique. Costs vary tremendously because of many factors including the distance between the sand source and the beach, the size of the nourishment, and the construction technique and timing. History indicates a ballpark range (with many exceptions) between about $½ to $5 million per mile for initial construction. Annual costs, considering renourishment needs, range from roughly $100,000 to $600,000 per mile per year. This works out to roughly $20 to $120 per year per foot of beachfront.

For example, the 2001 Gulf Shores, Alabama nourishment cost about 6 million dollars, covered about three miles and a renourishment interval of 7 to 10 years is predicted. Thus, the costs are about $2 million per mile for initial construction and, if renourishment costs are similar (they may be less), the annual cost will be roughly $250,000 per mile or $50 per front foot of beach. The Gulf Shores beach nourishment is funded by a local tax on rented bedrooms.

Searching for sand and studying the environmental impacts of nourishment can be expensive. For some beaches, these costs exceed the cost of construction. The size of the nourishment, including the amount of sand needed per mile of beach varies depending on the goals of the nourishment project and the existing erosion problem. For example, beach nourishments that include large sand dunes behind the beach require more sand.

The Ocean City, Maryland beach nourishment is probably one of the more expensive ones. It has over eight miles of beaches and a large sand dune. Renourishment was needed in 1992, 1994, and 1998 because of storm damage. The total cost has been roughly $60 million, which is an annual cost of about $110 per foot of beachfront.

These pictures show Myrtle Beach before (1984, above) and after (2001, below) the 1997 beach nourishment. The beach is much wider and higher. The seawall is now buried beneath a sand dune.

Is beach nourishment, at $20 to $120 per year per foot of beachfront, expensive? Well, expensive is a relative term. Relative to the total economic value of the beach, beach nourishment is not expensive. Beaches are the linchpins of the American tourism industry and tourism is one of our nation's largest industries employing millions of people and generating billions of dollars in payrolls and taxes. Relative to the value of the buildings immediately adjacent to the beach, beach nourishment is not expensive. Property values can exceed $500 million per mile ($100,000 per beachfront foot), more than one thousand times the annual nourishment cost. Relative to the protection it provides during storms, beach nourishment is not expensive. Even the very expensive beach nourishment at Ocean City Maryland has already paid for itself in reduced storm damages.

Based on all this, more and more Americans, including our elected officials, are coming to the same conclusion as Mayor Ann Kulchin of San Clemente, California that "we can't afford to <u>not</u> nourish our beaches." Relative to other public infrastructure investments, such as roads and sewers, beach nourishment is not expensive. New highways can cost over $100 million per mile. The reconstruction of one single

Newport Beach, California's beaches are nourished.

interchange, Interstate-95 and the Capital Beltway, is costing over $500 million. Relative to the total federal budget, beach nourishment is not expensive. According to Howard Marlowe, the "federal" government spends about $100 million per year on all its shore protection projects. This total includes some seawalls and does not include beach nourishments that are funded by local and state governments. This is less than one-one hundredth of 1% of the federal budget. Most of all, relative to the immeasurable social and emotional value sandy beaches provide millions of Americans, beach nourishment is not expensive.

Local governments fund many beach nourishments like Gulf Shores, Alabama did with a bed tax. A few beach nourishments, like Fisher Island, Florida, are funded entirely with private monies. Many beach nourishments are funded by a combination of local, state, and federal government sources like Ocean City, Maryland and Sea Bright, New Jersey. A few state governments have dedicated funding for beach nourishment. New Jersey and Florida have dedicated some state taxes for beaches. Several other states fund beach projects year-to-year and beach-by-beach. Federal funding requires some significant local or state matching funds, or cost sharing, by law. This is typically two-thirds federal and one-third non-federal money. However, this percentage split is always hotly debated in Washington, D.C. For years, the President, regardless of party, has tried to get Congress to stop all federal funding for beach nourishment.

Beach nourishment was called the "fleecing of America" by *NBC Nightly News* anchor Brian Williams. The Myrtle Beach nourishment was used as the example and was portrayed as expensive, futile and primarily for the benefit of wealthy beachfront homeowners. However, on July 4th, 2001 just one week before that *NBC Nightly News* newscast aired, over 100,000 Americans from all socio-economic backgrounds were enjoying the nourished beach at Myrtle Beach. Nourishment not only saved the beach from the fate of a seawall, it also improved the public access for working Americans to enjoy the much wider, sandy beach. It may also have paid for itself several times over in increased tax revenues and storm damage reduction. But mainly, nourishment saved the sandy beach at Myrtle Beach like it has saved many other great American beaches.

If beach nourishment can work here, it can probably work anywhere! There was no sandy beach or dune seaward of the Sea Bright and Monmouth Beach, New Jersey seawall for several decades before the 1994 beach nourishment. There was deep water at the base of the wall and waves hit the wall everyday. Now there is a wide beach and dune (photo from 2001).

Are nourished beaches as good as natural beaches?

Although a nourished beach is supposed to be like a natural beach, there are differences. Most of the differences occur because the nourishment sand does not perfectly match the native beach sands. Some nourished beaches "feel" harder than natural beaches. Some nourished beaches have larger pieces of rock or coral or shell than the native beach sands. Many have a higher percentage of shell-hash, broken up

seashells. Also, the distribution of these shells is often different on nourished beaches. Natural beaches have winnowed out all these different sizes over years, decades, centuries, and millennia. Nourished beaches, although they come in a hydraulic slurry, do not have the benefit of years of wave action in the surf and swash zone. Over time, the nourished beaches get closer and closer to looking like the native beaches. It usually only takes one or two cycles of sand moving out to the sand bar and then migrating back to the beach, a few storms, for much of the shell-hash to be winnowed out. But this occurs only on the outer edges of the new beaches. The sand in the dunes and at the back of the beach stays much as it came out of the pipe. Over time, wind, sun, storms, plants and animals, such as ghost crabs, can make these sands more natural looking and feeling. It shouldn't be surprising that we cannot make beaches as well as nature – nature has had more experience and time.

"Shell-hash," pieces of broken up seashells, often are mixed in with nourished beach sands.

The ecology of nourished beaches is important. Most of this chapter has focused on the physical characteristics of nourished beaches. There are many serious concerns about whether or not the plant and animal communities get reestablished on man-made beaches. The research on this is fairly limited and often contradictory. Sea turtle nesting is one area of concern that has a significant amount of research extending over 20 years. Beach nourishment projects are typically scheduled to avoid the turtle-nesting season and significant efforts are made to avoid dredging up slow moving turtles the rest of the year. Earlier concerns about beach compaction and temperature differences for nesting in new beaches have been partially allayed. Turtle nesting has gone up in some nourished areas because there are now sandy beaches where there once were seawalls.

It appears that some parts of the beach ecosystem reestablish fairly well in nourished beaches. This is dependent on the physical features of the new beach. If the new sand matches the native sand well, the ecological value seems to get restored better and sooner. The plants and animals that live in the upper few inches of sand come back within several weeks to months after construction. These are hardy, aggressive species that are used to storms reshaping the beach and so this should be expected.

The full ecology of the offshore, underwater sand source mining areas, which can be called scars, can take years to recover and may not ever go back to the way they were before. The upland portions of the nourishment area, the beach and dune, also can take years to recover. These areas also naturally take years to recover from large storms. Often, we help this recovery after nourishment by planting dune and beach vegetation. Unfortunately, such "recovery" efforts may not be the best approach for endangered piping plover habitat. Many wildlife biologists are concerned that nourished beaches, and the activity that comes with them, harm the plovers.

Other critical concerns relate to water turbidity near nourished beaches and the habitat being buried by the pumped sand. Many beaches have nearshore hardbottom or reef habitat very close to the shore. There is typically more of this along the coasts with the erosion problems because the erosion has removed the naturally thin veneer of sand. This extremely valuable habitat is avoided with nourishment or some mitigation is usually required. Seagrass beds are another nearshore habitat that is usually avoided in nourishment. However, one estimate is that over 100 acres of nearshore reefs and 35 acres of seagrass beds have been directly buried by beach nourishment in Florida since 1970.

Many beach nourishment projects are downdrift of a major sand thief. This 2001 nourishment at Cocoa Beach repaired erosion caused by the Port Canaveral harbor just several miles north.

We are already modifying nourishment designs for environmental reasons and they will become more and more important for future "designer beaches." We are still learning about how nourished beaches function ecologically. More research is needed towards understanding the short-term and long-term biological impacts of beach nourishment. It is possible that we may learn something about the ecosystem costs of nourishment that leads us away from its use as a response to erosion. But for now, a nourished beach is more natural than the most common realistic alternative in America of a seawall with little or no sandy beach.

Wilderness beaches

Some of our great American wilderness beaches are suffering the same ills as our urban beaches. And it appears we are going to have to save them the same way. For example, the north end of Assateague Island National Seashore is feeling the impact of human activities outside its borders. The jetties and boat channel at Ocean City Inlet cut off the "river of sand" that feeds Assateague. The erosion changed the island's ecology. The frequency of overwashing increased because of the sand starvation. The island naturally overwashes once every two to three years but now it overwashes up to 20 times per year!

The National Park Service faces a dilemma because of the link between barrier island habitats and overwash. They would rather let nature take its course but there is something very unnatural about the island now. In the words of Carl Zimmerman, the natural resource manager for the Seashore, "it was the intent of Congress in establishing Assateague Island National Seashore that the park provide a protected enclave for the complex plant and animal communities, both terrestrial and aquatic, which characterize the Mid-Atlantic Coast, and fully illustrate the natural processes of change which shape the coastal environment... The mission ... is to preserve these unique coastal resources and the natural ecosystem conditions and processes on which they depend,..."

Thus the National Park Service has a reason to try some coastal engineering – to try to make this seashore act more naturally. Beach nourishment is planned for 2002 and sand bypassing is planned for the next 30 years. This is to slow the dramatic recession and island rollover and allow some of the natural habitats to restore themselves. Thus, an island that is being destroyed by engineering must now rely on better sand management and coastal engineering for its survival in the modern world. This great wilderness beach may be saved the same ways the great urban beaches are being saved in America: backing off development, bypassing sand at inlets, and beach nourishment. These are the ways to manage beach erosion that work.

Conclusion

The Prescription for Saving America's Beaches

Sandy beaches are some of America's greatest places. They are our longest playgrounds and a big part of our economy. The lives of millions of Americans are touched by sandy beaches. The geology of America's beaches is very different from place to place but waves are the primary force shaping the coast everywhere. They drive sand along the beaches in a "river of sand" that moves both ways. Interruptions of this natural system are some of the primary causes of beach erosion.

Prescription for: _*Saving America's Beaches*_

1. Back-off

2. Bypass sand

3. Beach nourishment

select one or more, repeat as needed

*Dr. Douglass*

There are two common misconceptions about beaches. The first is that beach erosion is just a natural consequence of sea level rise and storms. The ugly truth is that much of America's beach erosion problem is man-made. We are damaging our beaches unknowingly by interrupting the movement of sand along and to our coasts. We have removed over a billion cubic yards of sand from America's beaches in the last century. This includes sand dredged at ship channels and dumped offshore, sand trapped by coastal structures that is no longer free to move along the beaches, and sand that no longer reaches the beaches of the Pacific coast because of river engineering. This is a tremendous amount of sand and is more than has naturally eroded from our beaches. Much of the man-made erosion can be stopped with better sand bypassing at the ship channels. We can have safe channels and great beaches: indeed we must. We just need to "bring back the beaches."

The second common misconception is that beach nourishment does not work. In fact, beach nourishment has saved many of America's most popular beaches. The list of nourished beaches includes Miami Beach, Santa Monica Bay, Myrtle Beach, Ocean City, and over 100 other beaches where people live and vacation. Beach nourishment has worked to save sandy beaches on all four of America's coasts: Atlantic, Pacific, Gulf of Mexico and Great Lakes.

The human impact on the sandy beaches is like a savings account balance. We "withdraw" sand by removing it from the beach system. We "deposit" sand with beach nourishment. Where the withdrawals exceed the deposits, we have beach erosion. Where the deposits exceed or match the withdrawals, we have sandy beaches. The sandy beaches of America need more deposits, like beach nourishment, and fewer withdrawals.

The prescription for saving America's beaches is clear. There are three parts to the prescription and some combination of the three parts is needed to save each eroding beach. One, 'back off' – setback construction wherever and whenever possible to provide a sandy buffer zone between the waves and the buildings. Two, sand 'bypassing' – we need to restore the movement of sand along and to our beaches. And three, when the beach is suffering too much from not paying attention to the first two parts of the prescription, 'beach nourishment' can restore many of our sandy beaches. All this is needed to ensure healthy, sandy beaches for future generations of Americans.

To Learn More About Beaches

● investigate and join one of the following organizations dedicated to the preservation of beaches:

☺ American Shore and Beach Preservation Association and its local chapters (www.asbpa.org)
☺ American Coastal Coalition (www.coastalcoalition.org)
☺ Surfrider Foundation (www.surfrider.org)
☺ Florida Shore and Beach Preservation Association (www.fsbpa.com)

● read another book about beaches. Here are a few suggestions:

The evolving coast by R.A. Davis, Jr. (Scientific American Library, 1994) A very clear explanation of the geology of the world's beaches with great pictures and graphics by one of America's premier coastal geologists.

Waves and beaches by W. Bascom (Anchor Doubleday Press, 1980) This out-of-print book is the classic on beach geology and waves. If you can find it, read it and keep it.

Against the tide: the battle for America's beaches by C. Dean (Columbia University Press, 1999). A very readable book by the science editor of the *New York Times* that discusses some of the same issues as the book you have in your hands but comes to a different conclusion concerning beach nourishment.

Beach Nourishment and Protection by the National Research Council (National Academy Press, 1995). Written, for laymen, by a nationwide committee of experts, this book evaluates beach nourishment and concludes that it can work to widen and restore beaches.

Surfriders: in search of the perfect wave by M. Warshaw (CollinsPublishers, 1997). Spectacular pictures of waves and a history of surfing by the former editor of Surfer magazine.

Great storms of the Jersey shore by L. Savadore and M.T. Buchholz (Down the Shore Publishing, Harvey Cedars, NJ, 1993). A fascinating glimpse of what barrier islands look like during and after storms. While this book covers New Jersey, the pictures look like the damage seen at many other places in America.

Florida beaches and *California beaches* by P. Puterbaugh and A. Bisbort (Foghorn Press, Santa Rosa, Cal., 1999 and 1999·2nd ed.) These are two fantastic travel guide-books, not beach science books. Advertised as the "only guide to the best places to eat, stay, swim, and play on every beach" in the state, they describe and rate, on a scale of 1 to 5, every single public access beach in these two great beach states!

The Pacific Northwest coast: living with the shores of Oregon and Washington by P. Komar (Duke University Press, 1998). An explanation of the science of the beaches of these two states written for laymen by one of America's premier coastal oceanographers.

The "Fine Print"

Acknowledgements

Dozens of people provided information and ideas for this book including: Dr. Robert G. Dean, Dr. Ashish J. Mehta, Dr. Daniel M. Hanes, Dr. Kevin R. Bodge, Dr. Albert E. Browder, Helen Tidwell, Cristopher G. Creed, Erik J. Olsen, Jason Engle, James MacMahan, Thomas J. Campbell, Donald E. Guy, Jr., Harry deButts, Dr. George Crozier, Dr. Douglas Haywick, Dr. Barry Costa-Pierce, Jay Tanski, Lt. Billy Oates, Aram V. Terchunian, Dr. Timothy W. Kana, Dr. Charles Shabica, Dr. Robert Wiegel, Kimberly K. McKenna, Sally S. Davenport, Dr. Reinhard E. Flick, Steven Aceti, Walter F. Crampton, Thomas R. Kendall, Pam Slater, Lesley C. Ewing, Peter Ravella, Michael P. Walther, Phillip Hinesley, Carl Ferraro, Bradley Gane, Bill Finch, Jacqueline Savitz, Kim Sterrett, Ann J. Kulchin, Eric N. Munoz, Orville T. Magoon, Dr. Susan Halsey, Dr. Karl Nordstrum, Dr. Stephen P. Leatherman, Lynn M. Bocamazo, Bradley H. Pickel, Russell H. Boudreau, Jeffrey C. Cole, Richard A. Davis, Jr., Dr. William McDougal; Tony McDonald, Dr. James R. Houston, Dr. William B. Stronge, Paden E. Woodruff, III, Robert Brantley, Roxanne Dow, Spencer M. Rogers, Jr., James O'Connell, Dr. Paul Gayes, Howard Marlowe, Al Devereaux, Robert W. Douglass, Dr. J. Richard Weggel, William Eiser, Christopher P. Jones, Dr. Clifford Truitt, Tracy Rice, Jeff Gebert, Dr. William Dally, Greg Bass, Dr. Cyril Galvin, Dr. Charles Dill, Michael Chrzastowski, Dr. Wendy Carey, Anthony P. Pratt, Chris Webb, Augustus T. Rambo, Jr., David "Chuck" Mesa, Carl Zimmerman, Craig B. Leidersdorf, Bernard Moore, Harry Shoudy, Chris Mack, Gregory Woodell, Virginia Barker, Stan Tait, Kate Gooderham, Stephen H. Higgins, Chuck Hoelle, Robert C. Foley, Tom Hutchings, George Kaminsky, Brian Flynn, with apologies to those I am forgetting to acknowledge. Travel for this research was supported by the Mississippi-Alabama Sea Grant and the Florida Sea Grant College Programs through projects R/PS-1-PD and PD-01-09, funded by National Oceanic & Atmospheric Administration grants NA86RG-0039 and NA76RG-0120, respectively. The University of South Alabama supported this research with a faculty service development award.

Photo Credits

p. 29:Green (1900)(note: reprinted by Pelican Press, 2000); 34:W. Crampton; 36:US Coast Guard; 44:K. Bodge; 46, 58(upper):Univ. of Florida Coastal Engineering Archives; 48, 49:J. Tanski; 64(lower):R. Weggel; 64(bottom):R. Dean; 65(both):T. Campbell

References

Preface:

Ewing, L. 1999. Conference overview: Sand Rights '99: bringing back the beaches. Conference proceedings edited by L. Ewing, O.T. Magoon, and S. Robertson, Am. Soc. of Civil Engineers, Reston, Virginia.

Chapter 1:

Houston, J.R. 2002. The economic value of beaches. 15th Nat. Conf. on Beach Preservation Technology. Biloxi, MS.

Jersey Shore Partnership. 2001. *The Jersey Shore ... a pleasure for all seasons.*

King, P. 1999. The fiscal impact of beaches in California. Public Res. Institute report, San Francisco State Univ.

King, P. and M.J. Potepan, 1997. The economic value of California's beaches, Public Res. Institute report, San Francisco State Univ.

Stronge, W. 2001. The economic value of our beaches and coastal properties. *Proc. of 14th Nat. Conf. on Beach Preservation Technology.*

Chapter 2:

Davis, R. A., Jr. 1994. *The evolving coast.* Scientific American Library.

Williams, S.J., K. Dodd and K.K. Gohn. 1990. Coasts in Crisis. U.S. Geological Survey circular 1075.

Chapter 3:

Bascom, W. 1980. *Waves and Beaches.* Doubleday Anchor Press.

California Office of Emergency Services. 1996. *Tsunami: how to survive the hazard on California's coast* brochure.

Dean, R.G., G.A. Armstrong and N. Sitar. 1984. California coastal erosion and storm damage during the winter of 1982-83. National Research Council. National Academy Press.

Engle, J. 2002. Formation of a rip current predictive index. 15th Nat. Conf. on Beach Preservation Technology. Biloxi, MS.

Green, N.C. 1900. *The story of the 1900 Galveston Hurricane.* (note: reprinted by Pelican Press in 2000)

Guy, D.E., Jr. 2001. Fieldtrip notes: Coastal Zone 2001, Cleveland, July 15.

Inman, D. 1967. *The beach: a river of sand,* 16mm movie. Encyclopedia Britannica Educational Corp.

Komar, P.D. 1988. *Beach processes and sedimentation.* Prentice-Hall.

Surfing magazine. Feb. 2002 issue.

U.S. Army Engineers. 1984. Shore Protection Manual.

della Cava, M.R. 2001. Sharky and full of death. *USA Today.* 2/20/01

Chapter 4:

Bodge, K.R. 1998. Sediment Management at Inlets/Harbors. Draft of Coastal Engineering Manual chapter.

Bodge, K.R. 1994. Port Canaveral inlet management plan report by Olsen Assoc. for Canaveral Port Authority

Carey, W., E.M. Maurmeyer, and A.P. Pratt, ed. 2001. Coast of Maryland and Delaware Field Trip Guidebook: 2001 Am. Shore and Beach Pres. Assoc. Conf. May 13, 2001.

Daley, W., C. Jones, T. Mootoo, A. Terchunian and G. Vegliante. 2000. A blueprint for coastal management: the West Hampton Dunes story. *Shore & Beach,* 68.1, Jan.

Dean, R.G. 1988. Managing sand and preserving shorelines. *Oceanus,* 30:3:49-55.

Dean, R.G. and M.P. O'Brien. 1987. Florida's east coast inlets: shoreline effects and recommended action. UFL/COEL-87-017.

Dean, R.G. and M.P. O'Brien. 1987. Florida's west coast inlets: shoreline effects and recommended action. UFL/COEL-87-018.

Griggs, G.B. 2001. California's beaches: lessons from the past and recommendations for the future. abstract. California Shore & Beach Preservation Assoc. meeting. Nov.

Guy, D.E., Jr. 2001. Fieldtrip notes: Coastal Zone 2001, Cleveland, July 15.

Kaminsky, G.M., M.C. Buijsman, and P. Ruggiero. 2000. Predicting shoreline change at decadal scale in the Pacific Northwest, USA. *Proc. Coastal Engineering 2000*, Am. Soc. of Civil Engrs., pp. 2400-2413.

Kendall, T.R. , J.C. Vick and L.M. Forsman. 1991. Sand as a resource: managing and mining the northern California coast. *The California Coastal Zone Experience*, Domurant, et al. ed., Am. Soc. of Civil Engrs.

Komar, P.D. 2000. Coastal erosion – underlying factors and human impacts. *Shore & Beach*, 68:1, Jan.

Magoon, O.T. and B.L. Edge. 1998. Sand rights – the fragile coastal balance. *Proc. emerging trends in beach erosion and sand rights law*, June 3-5.

Malone, S. 2001. Pennsylvania's use of interstate consistency to condition a federal dredging activity in Conneaut Harbor, Ohio. *Proc. Coastal Zone 2001*.

Marino, A.J and A.J. Mehta. 1986. Sediment trapping at Florida's east coast tidal inlets. UFL-COEL-86/006.

Raichle, A.W., K.R. Bodge and E.J. Olsen. 1997. St. Mary's entrance inlet management study, report by Olsen Assoc. for Nassau Soil and Water Cons. District.

Runyan, K. 2001. Contributions of coastal cliff erosion to the beach sand budget in California and the effects of armoring. abstract. California Shore & Beach Preservation Assoc. meeting. Nov.

Chapter 5:

Beachler, K. 1994. The positive impacts to neighboring beaches from the Delray Beach nourishment program. *Proc. 1994 Nat. Conf. on Beach Preservation Technology*.

Bocamazo, L.M., D. Rahoy, and W.G. Grosskopf. 2001. Atlantic coast of New Jersey, Sea Bright to Manasquan post-construction monitoring summary. Joint conference of Am. Shore and Beach Preservation Assoc. and the Am. Coastal Coalition. Washington, DC, May 14.

Bodge, K. 2000. Beach nourishment engineering. Bays, bayous and beaches symposium. Mobile, AL, April.

Bodge, K. 1991. Beach nourishment with aragonite and tuned structures. *Proc. Coastal Engineering Practice Conf.*, Am. Soc. of Civil Engineers. p. 73.

Campbell, T. 2001. Successful beach nourishments in Florida. Joint conference of Am. Shore and Beach Preservation Assoc. and the Am. Coastal Coalition. Washington, DC, May 14.

Carey, W., E.M. Maurmeyer, and A.P. Pratt, ed. 2001. Coast of Maryland and Delaware Field Trip Guidebook: 2001 Am. Shore and Beach Pres. Assoc. Conf. May 13, 2001.

Chrzastowski, M. 1991. The building, deterioration and proposed rebuilding of the Chicago lakefront. *Shore & Beach*. 59:2:Apr.

Dean, R.G. 2001. Where has all the beach nourishment sand gone? Florida Shore & Beach Preservation Assoc. meeting. Duck Key, FL, Sept. 2001.

Dean, R. 2001. Partial notes for EOC 6934:beach nourishment: theory and applications.

Dornhelm, R.B. 1995. The Coney Island public beach and boardwalk improvement of 1923. *Shore & Beach*. 63:1:Jan.

Flick, R.E. 1993. The myth and reality of southern California beaches. *Shore & Beach*. 61:3:Jul.

Flynn, B. 2002. The landmark Miami Beach project. 15th Nat. Conf. on Beach Preservation Technology. Biloxi, MS.

Kana, T.W., R.E. Katamarian and P.A. McKee. 1997. The 1986-1995 Myrtle Beach nourishment project ten-year performance summary. *Shore & Beach*. 65:1:Jan.

Leidersdorf, C.B., Hollar, R.C. and G. Woodell. 1994. Human intervention with the beaches of Santa Monica Bay, California. *Shore & Beach*. 62:3:Jul.

Lindeman, K. and 69 others. 2000. 70 Ph.D. scientists urge higher environmental standards in beach dredge and fill projects. letter from Environmental Defense to U. S. Army Engineer District, Jacksonville, June 27.

Olsen, E. 2000. talk on Wilmington Harbor, NC project. Florida Shore & Beach Pres. Assoc. meeting. Captiva Island, FL. Sept. 13.

National Research Council. 1995. *Beach Nourishment and Protection*. National Academy Press.

Ray, G.L. 2001. Responses of benthic invertebrate assemblages to the Asbury-Manasquan Inlet beach nourishment project. Coastal ecosystems and federal activities technical training symposium. US Fish & Wildlife Service. Gulf Shores, AL, Aug. 21.

Rogers, S.M. 2001. Beach nourishment for hurricane protection: North Carolina project performance in Hurricanes Dennis and Floyd. *Proc. 2001 Nat. Conf. on Beach Preservation Technology*.

Sasso, H. 1997. Dade County regional sediment budget. Coastal Systems International report for Metro. Dade County.

U.S. Army Engineers, Baltimore District. undated brochure. Beach replenishment and hurricane protection project: Ocean City, Maryland.

Weaver, R. 2002. The 26-mile shore protection project at Harrison County, MS. 15th Nat. Conf. on Beach Preservation Technology. Biloxi, MS.

Index